Sustainable Fashion and Textile Recycling

Sustainable Fashion and Textile Recycling

Editors

Hanna de la Motte
Åsa Östlund

MDPI • Basel • Beijing • Wuhan • Barcelona • Belgrade • Manchester • Tokyo • Cluj • Tianjin

Editors
Hanna de la Motte
RISE Research Institutes of Sweden
Sweden

Åsa Östlund
Tree to Textile
Sweden

Editorial Office
MDPI
St. Alban-Anlage 66
4052 Basel, Switzerland

This is a reprint of articles from the Special Issue published online in the open access journal *Sustainability* (ISSN 2071-1050) (available at: https://www.mdpi.com/journal/sustainability/special_issues/sustainable_fashion).

For citation purposes, cite each article independently as indicated on the article page online and as indicated below:

LastName, A.A.; LastName, B.B.; LastName, C.C. Article Title. *Journal Name* **Year**, *Volume Number*, Page Range.

ISBN 978-3-0365-5787-8 (Hbk)
ISBN 978-3-0365-5788-5 (PDF)

© 2022 by the authors. Articles in this book are Open Access and distributed under the Creative Commons Attribution (CC BY) license, which allows users to download, copy and build upon published articles, as long as the author and publisher are properly credited, which ensures maximum dissemination and a wider impact of our publications.

The book as a whole is distributed by MDPI under the terms and conditions of the Creative Commons license CC BY-NC-ND.

Contents

About the Editors vii

Hannah Auerbach George, Marie Stenton, Veronika Kapsali, Richard S. Blackburn and Joseph A. Houghton
Referencing Historical Practices and Emergent Technologies in the Future Development of Sustainable Textiles: A Case Study Exploring "Ardil", a UK-Based Regenerated Protein Fibre
Reprinted from: *Sustainability* 2022, *14*, 8414, doi:10.3390/su14148414 1

Katarina Lindström, Therese Sjöblom, Anders Persson and Nawar Kadi
Improving Mechanical Textile Recycling by Lubricant Pre-Treatment to Mitigate Length Loss of Fibers
Reprinted from: *Sustainability* 2020, *12*, 8706, doi:10.3390/su12208706 23

Ajinkya Powar, Anne Perwuelz, Nemeshwaree Behary, Le Vinh Hoang, Aussenac Thierry, Carmen Loghin, Stelian Sergiu Maier, Jinping Guan and Guoqiang Chen
Environmental Profile Study of Ozone Decolorization of Reactive Dyed Cotton Textiles by Utilizing Life Cycle Assessment
Reprinted from: *Sustainability* 2021, *13*, 1225, doi:10.3390/su13031225 37

Miriam Ribul, Kate Goldsworthy and Carole Collet
Material-Driven Textile Design (MDTD): A Methodology for Designing Circular Material-Driven Fabrication and Finishing Processes in the Materials Science Laboratory
Reprinted from: *Sustainability* 2021, *13*, 1268, doi:10.3390/su13031268 49

Felice Diekel, Natalia Mikosch, Vanessa Bach and Matthias Finkbeiner
Life Cycle Based Comparison of Textile Ecolabels
Reprinted from: *Sustainability* 2021, *13*, 1751, doi:10.3390/su13041751 67

Paulien Harmsen, Michiel Scheffer and Harriette Bos
Textiles for Circular Fashion: The Logic behind Recycling Options
Reprinted from: *Sustainability* 2021, *13*, 9714, doi:10.3390/su13179714 91

Joséphine Riemens, Andrée-Anne Lemieux, Samir Lamouri and Léonore Garnier
A Delphi-Régnier Study Addressing the Challenges of Textile Recycling in Europe for the Fashion and Apparel Industry
Reprinted from: *Sustainability* 2021, *13*, 11700, doi:10.3390/su132111700 109

Seonju Kam
Three-Dimensional Printing Fashion Product Design with Emotional Durability Based on Korean Aesthetics
Reprinted from: *Sustainability* 2022, *14*, 240, doi:10.3390/su14010240 139

Jenny Bengtsson, Anna Peterson, Alexander Idström, Hanna de la Motte and Kerstin Jedvert
Chemical Recycling of a Textile Blend from Polyester and Viscose, Part II: Mechanism and Reactivity during Alkaline Hydrolysis of Textile Polyester
Reprinted from: *Sustainability* 2022, *14*, 6911, doi:10.3390/su14116911 153

Anna Peterson, Johan Wallinder, Jenny Bengtsson, Alexander Idström, Marta Bialik, Kerstin Jedvert and Hanna de la Motte
Chemical Recycling of a Textile Blend from Polyester and Viscose, Part I: Process Description, Characterization, and Utilization of the Recycled Cellulose
Reprinted from: *Sustainability* 2022, *14*, 7272, doi:10.3390/su14127272 163

Hafeezullah Memon, Henock Solomon Ayele, Hanur Meku Yesuf and Li Sun
Investigation of the Physical Properties of Yarn Produced from Textile Waste by Optimizing Their Proportions
Reprinted from: *Sustainability* **2022**, *14*, 9453, doi:10.3390/su14159453 . . . . **179**

About the Editors

Hanna de la Motte

Dr. Hanna de la Motte is VP of the Business and Innovation Area: Material Transition at RISE (Research Institutes of Sweden), with 15 years of knowledge on materials science and sustainability. Hanna specializes in cellulose science, fiber development, and chemical recycling of textiles and plastics, and her work has led to scalable innovations in Sweden. She is passionate about the material transition and working through the entire value chain, from production and sustainable business models to end-of-life issues as well as traceability and quality assurance. Another important focus for her is the collaboration between different actors from a systems perspective to create the sustainable and scalable material ecosystem of the future.

Åsa Östlund

Dr. Åsa Östlund is the Head of R&D at Tree to Textile and has been working within the research area of sustainability in textiles and fashion for over 10 years. Åsa is an expert in the fields of 'dissolution and regeneration of cellulose' and 'textile recycling'. She also has experience in driving cross-disciplinary research and development and has previously led larger initiatives such as the research program Mistra Future Fashion (2011–2019), the European network COST FP 1205 Innovative applications of regenerated wood cellulose fibers (2012–2016), among others.

Review

Referencing Historical Practices and Emergent Technologies in the Future Development of Sustainable Textiles: A Case Study Exploring "Ardil", a UK-Based Regenerated Protein Fibre

Hannah Auerbach George [1], Marie Stenton [2], Veronika Kapsali [2], Richard S. Blackburn [3] and Joseph A. Houghton [3,*]

1. Victoria and Albert Museum, Cromwell Rd., London SW7 2RL, UK; h.auerbachgeorge@vam.ac.uk
2. London College of Fashion, University of the Arts London, London SW1P 4JU, UK; m.stenton0620191@fashion.arts.ac.uk (M.S.); veronika.kapsali@fashion.arts.ac.uk (V.K.)
3. Leeds Institute of Textiles and Colour, University of Leeds, Leeds LS2 9JT, UK; r.s.blackburn@leeds.ac.uk
* Correspondence: j.a.houghton@leeds.ac.uk

Citation: Auerbach George, H.; Stenton, M.; Kapsali, V.; Blackburn, R.S.; Houghton, J.A. Referencing Historical Practices and Emergent Technologies in the Future Development of Sustainable Textiles: A Case Study Exploring "Ardil", a UK-Based Regenerated Protein Fibre. *Sustainability* 2022, 14, 8414. https://doi.org/10.3390/su14148414

Academic Editors: Hanna de la Motte and Asa Ostlund

Received: 24 May 2022
Accepted: 6 July 2022
Published: 9 July 2022

Publisher's Note: MDPI stays neutral with regard to jurisdictional claims in published maps and institutional affiliations.

Copyright: © 2022 by the authors. Licensee MDPI, Basel, Switzerland. This article is an open access article distributed under the terms and conditions of the Creative Commons Attribution (CC BY) license (https://creativecommons.org/licenses/by/4.0/).

Abstract: We are currently experiencing a global environmental crisis. Our waste culture is leading to huge irreversible damage to our planet and ecosystems. This is particularly evident in both the textile and food sectors, with a system-wide restructuring as to how we consume and source materials becoming ever more urgent. By considering our waste as resource, we can access a vast source of raw materials that is now being recognised as such. Viable materials in the form of waste have the potential for conversion into textiles. However, this proposed solution to our contemporary crisis is not new technology. Throughout the 20th century, science and industry have researched and developed materials from food waste to meet global demand for textiles in times of need, with a major development during the world wars being the invention of regenerated protein fibres (RPFs). For various reasons, this research was abandoned, but much of the development work remains valid. This research critically analyses work that has previously been done in the sector to better our understanding of the historical hindrances to the progression of this technology. By applying modern thinking and scientific advances to historical challenges, there is the potential to overcome previous barriers to utilising food waste as a resource. One of the key influences in the discontinuation of RPFs was the rise of petrochemical textiles. Our current understanding of the detriment caused by petrochemicals warrants a further review of historical emergent technologies. This paper uses Ardil fibre as a case study, and shows that there is a clear disparity between the location of historic research and where the research would now be helpful. Ardil was a British-made product, using peanuts sourced from the British Empire as the source of protein. Techniques used in the processing of Ardil could be better utilised by countries and climates currently producing large amounts of peanut byproducts and waste. Through this research, another historical concern that thwarted Ardil's acceptance as a mainstream fibre was discovered to be its poor tensile strength. However, contemporary garment life cycles are far shorter than historical ones, with built-in obsolescence now being considered as a solution to fast fashion cycles by matching the longevity of the fibre to the expected use phase of the garment, but ensuring suitable disposal methods, such as composting. This research highlights the need for cross-disciplinary collaboration between sectors, with a specific focus on the wealth of valuable information available within historical archives for modern sustainability goals.

Keywords: regenerated fibres; regenerated protein fibres; waste; circular economy; valorisation; garment industry; manmade fibres; textile processing; textile history; Ardil

1. Introduction

We are firmly in a period of environmental crisis, with humanity's impact on the planet becoming ever more evident, and the time before irreversible damage has been caused is rapidly dwindling. Two of the largest sectors contributing to this environmental crisis are the textile and food production industries. The size of the textiles sector has been increasing steadily in recent years, and the garment sector alone is estimated to produce 3.3 billion tonnes of CO_2e annually [1]. Globally, fibre production reached 109 million tonnes (Mt) in 2020, representing a 10-fold increase since 1950, of which synthetic fibres represent roughly 62% of production mass. Global fibre production is expected to increase by another 34% to 146 Mt by 2030. Worldwide fibre production per person increased from 8.4 kg per person in 1975 to 14 kg per person in 2020 [2]. With the global population simultaneously expected to increase by 40% by 2050, the environmental burden of the textile industry will increase rapidly in the coming decades [3]. The increase in clothing purchased can be attributed to several different factors, such as the comparably slow rise in clothing prices when compared to other consumer goods [4], and the advent of the phenomenon known as "fast fashion" [5,6]. With the turnover of fashion being so high, and the cost of clothing being comparably low, the amount of waste generated per annum is vastly increasing with this new model of textile consumerism.

Understanding the environmental impact of different textiles is complex, and the literature often provides conflicting data. For example, Moazzem et al. claim that a key contributing factor to the environmental impact of the textile supply chain is the consumer use phase in garment life cycles [7]. However, a 2017 report from WRAP cites the fibre production stage as the highest contributor to the carbon footprint of clothing, with roughly 11 Mt of CO_2e produced in 2016 (out of a total of 26.2 Mt of CO_2e across the whole supply chain) [8]. These conflicts often arise because the process of measuring and recording the environmental impact of a garment throughout its life cycle is highly complex, making any comparisons across fibre type subject to debate. Recently, a widely used tool for mapping the sustainability of fibre types and materials—the Higg Materials Sustainability Index (MSI)—has come under scrutiny, as it only considers the production phase in its assessment. [9].

The environmental issues and concerns with synthetic fibres are well documented, and have dominated the media headlines in recent years. It is often assumed that natural fibres are more sustainable when compared to synthetics. However, the environmental effect of the production of natural fibres cannot be ignored: the recent increase in the desirability of natural fibres in an attempt to be environmentally conscious has been shown to cause myriad issues. Cotton fibre is a vast industry estimated to employ nearly 7% of labour in developing countries [10]. Operating the cotton industry at this scale comes at a huge detriment to the environment. Cotton is very difficult to grow successfully, requiring huge amounts of water and pesticides to sustain the crop, which occupies only 2.4% of cultivated land but consumes 6% of pesticides and 16% of insecticides globally [11]. Production of cotton uses vast amounts of water; 69% of all water used in fibre production is attributed to cotton, with 1 kg of finished fabric taking about 20,000 litres to produce [12].

Protein fibres also have issues. Farming and harvesting silk also involves intensive and environmentally harmful processes, and is ranked higher than most other fibres on the Higg MSI [13]. Although silk fibre represents a much smaller proportion of the global fibre market than cotton, at just 0.1% of the global market, the value of silk was expected to be USD 16.94 billion in 2021 due to increased desire for this fibre [2]. Wool fibre production has an incredibly high carbon footprint, accounting for 36% of the total carbon footprint in fibre production for clothing in use in the UK in 2009 [14], and the desertification of Mongolia has been attributed to the farming of cashmere goats [15,16]. While, again, these statistics only consider the production phase of these fibres, this must be understood in the context of the renewed interest in natural fibres and the vast increase in demand. One hundred years ago, the quantity of textiles required to meet global demand was far less than today, due to the lower population and the different approach to clothing. A coat lasted a lifetime, as opposed to a fashion season, which meant that the pressure on the production of natural fibres was lower, and the associated environmental impact was limited. The synergistic effects of the increased demand for natural "sustainable" fibres from a more environmentally conscious consumer base, along with the evidenced increase in environmental impact resulting from overstrained natural fibre production routes, opens up an exciting opportunity for novel research into potential feedstocks for sustainable fibre production. However, there is potential not only in exploring "new" processes, but also in learning from historical precedent to draw inspiration for a more sustainable future.

There is a genuine need for sustainable fibres in contemporary society; however, the problems of overpopulation and lack of raw materials are not new. This was acutely felt during the first half of the 20th century, with the economic pressure created by two world wars and the growing global population. During the interwar period, in response to this hardship and strain on textile resources, government hopes turned to an emerging science—regenerated protein fibres (RPFs). Their revolutionary idea was to use regenerated protein fibres created in a lab to reduce their reliance on natural fibres grown in fields or on animals. In theory, RPFs could be made from an array of protein sources more simply and economically than traditional fibres such as wool, which requires farming livestock [17]. It was hoped that these new regenerated fibres would provide an economical and competitive alternative to natural fibre resources [18], and across Europe many companies began experimenting with RPFs on a commercial scale, in the hope that these fibres would be the future of the textile industry [19]. While there are some examples of contemporary RPFs in today's market, such as QMilch, this paper focuses on historical examples of RPFs.

The process developed during the interwar period to produce RPFs was as follows [20]:

1. Dissolution of the protein in a suitable solvent. This is typically an aqueous solution of a diluted alkali.
2. Denaturing and unfolding the protein molecules into a linear state. In the case of casein this is achieved via the introduction of NaOH which, when given enough time, breaks the proteins' secondary structure. This enables extrusion through small holes within a spinneret.
3. Extrusion of this protein "dope" through a spinneret directly into a coagulation bath in a wet spinning process. This involves controlled precipitation of the protein in the coagulation bath, forming continuous filaments. The coagulation bath usually consists of an anti-solvent—typically an aqueous solution of diluted acid—and other chemical additives. Salt is added to increase the osmotic pressure within the bath, causing fibre shrinkage to aid with protein molecule orientation and to prevent fibres from sticking together. A crosslinking agent (most commonly formaldehyde) is added to improve the tensile properties of the fibre. The filaments can then undergo further mechanical and chemical processes to increase their functional properties, including chemical hardening in a separate bath. The most common chemicals used historically for RPFs are sulphuric acid, sodium sulphate, magnesium sulphate, and formaldehyde.
4. Drawing filaments. A mechanical process of gradually stretching the fibre to aid in protein molecule alignment and increase fibre crystallinity, leading to an increase in tensile strength.

Despite showing early promise, this industry died out after World War II. Some of the key factors in the disappearance of these fibres included problems with tensile strength—particularly wet strength—and the availability of raw materials. RPFs were a revolutionary idea that promised to solve issues of supply and demand, but ultimately failed. However, this method of creating fibres has been shown to have potential in the realm of sustainable textiles, as the protein used can be upcycled from waste. There is renewed interest in this technology, as the need for more sustainable fibres grows ever stronger. It is also possible for regenerated fibres to be broken down post-use, creating a material with a circular economy. At present, further research into environmentally friendly methods of producing RPFs is required, as historically the process of crosslinking proteins uses harmful chemicals (e.g., formaldehyde). In particular, systems supporting the life cycle of these fibres, such as waste collection and composting of the garments, have previously been identified as key areas for further research [21]. This paper explores the reasons for the disappearance of RPFs, and asks what lessons can be applied to sustainable contemporary textile design. By reappraising historical examples of RPFs against contemporary scientific developments, it is hoped that a potential future sustainable method can be demonstrated.

Ardil fibre was chosen as a case study, with the aim of understanding the shift away from these promising early fibre developments and the potential lessons to be gained from past RPF research. Ardil is the brand name of a fibre first developed in the 1930s by the British company Imperial Chemical Industries (ICI). It was made from groundnuts, more commonly known as and hereafter referred to as peanuts (although the literature uses the names interchangeably). Ardil's development coincided with periods of shortage and austerity during the world wars—a time when manufacturers were seeking alternatives to natural fibres. During these wars, textiles were deemed to be of vital importance, with wool for uniforms being considered just as important as bullets [22]. Not only that, but the strain of the war effort on the textile industry was immense, with 65% of the production capacity being diverted to the creation of government fabrics [23]. This vastly reduced the available fibres and fabrics for the general consumer. There was also fear of disruption of the wool trade routes leading to material shortages. All of these factors combined resulted in the heavy investment in and rapid development of fibres that could be produced from alternative feedstocks, such as Ardil. However, Ardil production did not continue beyond the 1950s. By closely scrutinising the development, success, and subsequent downfall of this once highly promising RPF, and by employing modern techniques to critically analyse the sustainability of the process, we can begin a holistic approach to the development of contemporary RPFs.

2. Materials and Methods

This paper combines literature from design, science, and industry to investigate how textile materials can be created from food waste to elevate pressure on finite resources for textiles. The authors used Ardil as a specific example of an RPF as a case study, as it was one of the few examples of a British RPF to be put into commercial production with moderate success, providing useful insights for contemporary textiles. The authors first planned the paper, identifying Ardil as an appropriate case study before outlining the structure of the paper. Next, data were collected using literature searches and by identifying and visiting appropriate collections. This information was then critically analysed by the authors before reporting their findings and opinions on the subject. A detailed review of the RPF patents was used in order to understand the chronological progression of the technology. Using historical sources including newspaper articles, magazines, journals, and archival collections, this paper first outlines the story of Ardil, seeking to understand the nuances behind its development and, for the first time, the reasons for its subsequent failure. Secondly, this is contextualised through a contemporary lens, questioning what knowledge can be applied to the ongoing research and development of contemporary RPFs. A timeline illustrating the rise and fall of Ardil helps explain its place in history and the simultaneous shifts in the global textile market and the wider economy (Figure 1). Three main areas in the story of Ardil are considered:

1. Development of the historical technology;
2. Marketing and public reception of Ardil;
3. Social, political, and economic factors.

This research has been challenging, as there are very few recorded examples of Ardil in contemporary museum collections today. Brooks argues this may be a result of Ardil's poor longevity as a fibre and the difficulty in distinguishing fabrics made from Ardil from other fibres, such as wool [24]. As such, unless there is clear supporting evidence or labelling, many Ardil fabrics may sit in collections unnoticed. The lack of known examples of Ardil and the rarity of academic literature relating to it have previously hindered in-depth research into this fibre, as the material is not easily accessible. Whilst conducting the archival research for this paper, several little-known or previously unrecorded examples of Ardil were discovered. These discoveries have helped to shape further understanding of the prevalence and historical importance of this fibre.

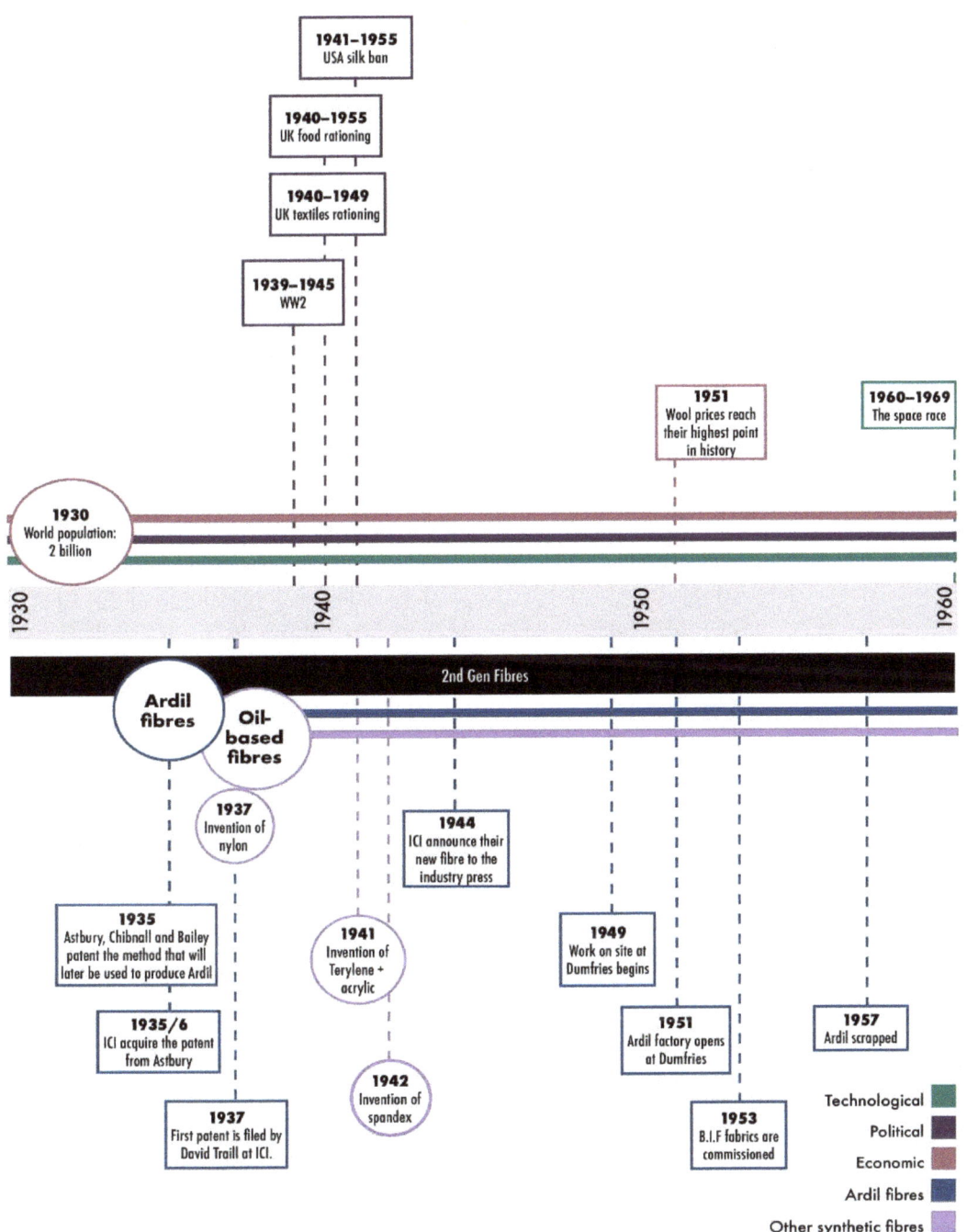

Figure 1. This timeline shows the key events in the development, marketing, and subsequent demise of Ardil fibre, as well as the social and economic events that influenced its trajectory.

3. Results

3.1. Historical Background

3.1.1. RPF Background

The development of RPFs stretches back to the 19th century, with the first known example being a fibre made from gelatine by the scientist Adam Millar in 1894. Following on from Millar's work, other scientists developed RPFs using sources of protein that included maize, soya beans, chicken feathers, albumin, and eggs. It was not until the 1930s that the technology sufficiently advanced to become scalable. In 1935, the Italian scientist Antonio Ferretti created a commercially viable casein fibre, which was marketed as Lanital. The Italians were so proud of their new fibre technology that it was used in the manufacture of Italian army uniforms, under Mussolini's rule. However, the army were not impressed by the poor quality of Lanital as compared to wool [25]. Lanital, like the majority of RPFs developed during the interwar period, proved unsuccessful, as it could not compete with natural fibres such as cotton and wool [26]. Although RPFs successfully mimicked desirable properties of natural fibres such as soft handle, they were easily damaged and became weak when damp [27].

3.1.2. Imperial Chemical Industries Background

ICI was one of the largest science and technology companies of the 20th century. They were responsible for creating well-known products such as Perspex, Polythene, Dulux paints, and hundreds more. Though predominantly a chemical company, ICI were acutely aware of the shifts within textiles, and of the potential role their chemical expertise could play in the sector. In the early 1900s, it was widely believed that the future of the textile industry lay in scientific development and the exciting properties that technology could imbue. Alongside RPFs, this was also the age of development for petrochemical-derived fibres such as nylon and polyester, which would eventually supersede RPFs to become the prevalent materials they are today, but for a time both technologies were equally feted. The work of ICI was a clear example of this, with involvement in developing several manmade fibres, including synthetic fibres from the petrochemical industry as well as RPFs [28].

3.2. Development of the Historical Technology

3.2.1. Invention of the Process

Though ICI were responsible for the development, marketing, and even naming of Ardil, the science behind the fibre originated from the work of William Astbury at The University of Leeds. Astbury worked in textile physics at Leeds from 1928, and held the Chair of Professor of Biomolecular Structure from 1946 until his death in 1961 [29]. He focused much of his work on using X-rays to study biological molecules—particularly wool and hair fibres. His work with Florence Bell on X-ray crystallography of biological molecules revealed the regular, ordered structure of DNA, which laid the foundations for the structural identification of DNA by Crick and Watson in 1953.

In addition to this work, in 1935 Astbury developed a technique for producing fibres from vegetable proteins with his colleagues Chibnall and Bailey at Imperial College London [30]. Their technique involved denaturing globular proteins—a process of chemically unfurling the molecules of protein in order to "refold" it into a fibrous form. These globular proteins are found in different sources of vegetable protein, including hemp seeds and peanuts.

3.2.2. The Role of ICI

To understand why ICI were so interested in Astbury's work, it is important to understand where the technology was at this point in time. Analysis of patents filed relating to RPFs, from their invention in the 1870s onwards, shows that prior to Astbury's work RPFs were more commonly produced from animal proteins. Before Astbury and his colleagues filed their patent in October 1935, only three other researchers had filed patents for RPFs since Adam Millar in 1894. These were Dr Friedrich Todtenhaupt, Herman Timpe, and Antonio Ferretti, whose methods all focused predominantly on milk casein, gelatine, eggs, and albumin. In Astbury's patent, he specifies "Production of silk and wool-like threads from 'vegetable' proteins belonging to globulin group" [31]. This patent would have likely been of great interest to ICI, as it specifically worked with seeds and legumes, as vegetable proteins were in higher abundance throughout the world [26]. Importantly, using vegetable proteins would also cut out the additional and costly step of farming livestock that the previous patents for animal-protein-derived RPFs used.

Quickly realising the potential in Astbury's research, ICI formally purchased the patents the following year [32]. The responsibility of further developing Ardil for commercial use was handed to David Traill in ICI's Nobel Division [33], named after Alfred Nobel, who originally founded the factory at this site [34]. Traill set to work finding the most suitable source of protein to use with Astbury's methods. In patents filed by Traill between 1937 and 1939, he specifies peanuts as well as casein, until the 1940s onwards, where he focuses purely on peanuts [35,36]. Traill's experiments had found peanut protein to be the most suited to his purpose [37]. Peanuts were commonly imported to the UK as a foodstuff, and so could be relied upon as a regular source of protein to work with; the Gambia alone imported 10,275.6 tonnes to the UK in 1935 [38]. The main product produced from peanuts at this time was peanut oil, with the leftover meal used as animal feed or considered waste. As discussed, there were already several examples of casein regenerated protein fibres in commercial production [39], but none using peanut fibre yet (the only other historical example of a peanut fibre being developed commercially was Sarelon in the US, where peanuts can be grown [40,41]). With the parameters set for their new product, Ardil fibre was christened after the location where Traill worked—at an ICI plant in Ardrossan on the Ardeer Peninsula in Ayrshire.

3.2.3. The Chemical Process

The chemical process of creating Ardil was described in 1955 [19]: Peanuts are first crushed to extract the oil used as an ingredient in food manufacture (e.g., margarine) and personal care products (e.g., soap). The residue consists of approximately equal parts of carbohydrate and protein; the protein is extracted, and is washed and dried to obtain a white powder called Ardein, which is primarily composed of a protein called arachin. Ardein is dissolved in caustic soda and extruded through spinnerets into a coagulation bath, followed by hardening, washing, crimping, and cutting into staple fibres ready for spinning (Figure 2a). As discussed in the introduction, the hardening stage often employed the use of formaldehyde as a crosslinking agent; this posed serious environmental concerns with regards to wastewater. This, in combination with the fact that wet spinning as a process is very water-intensive, with large amounts of water being required for every stage of the process, means that even though this protein fibre utilised a "waste" feedstock, the actual environmental impact of the process was likely to be relatively harmful. The undissolved part of the peanut meal is recovered, and is valuable cattle food [42].

As Ardil was not produced as a continuous filament, it was very versatile for spinning into a variety of yarn weights. The fibres could be made into several different deniers to suit garments or interior textiles accordingly. In order to promote and explain their new fibre to potential customers, ICI produced a manual for manufacturers [26]. The manual explains that there were three main types: B, F, and K, each with slightly different properties and subsequent applications. Another important distinction between the types was the shade; Ardil was naturally a fawn colour, with the manual suggesting that to achieve a pure white shade, bleaching was required. This may have been seen as drawback by some manufacturers, as it would affect the shades that could be achieved through dyeing. The reasons for producing these different product specifications for Ardil are not immediately obvious, but the unusual product categories imply that ICI were compensating for shortcomings of Ardil.

(a) (b)

Figure 2. (a) "The moment when 'Ardil' first becomes a fibre": Filaments of Ardil are drawn from fibres in the coagulating bath; Catalyst Science Discovery Centre and Museum. (b) "Ardil Protein Fibre Factory, Dumfries" by Henry Rushbury, with the Ardil Tower visible; Catalyst Science Discovery Centre and Museum.

3.2.4. Blends with Ardil

Despite high expectations, during initial testing, it became clear that Ardil was not particularly strong as a fibre; it had poor tensile strength when wet compared to other fibres, such as wool (Table 1) [26].

Table 1. Comparison of the principal physical properties of "Ardil" fibre and wool.

Property	"Ardil" Fibre	Wool
Specific gravity	1.31	1.31
Tensile strength (kg./sq.mm.)	8–10	12–20
Elongation at break (%)	40–60	30

It was therefore decided that Ardil would perform better when blended with wool and other fibres, such as rayon, which would stabilise it. The most notable incentive for using Ardil as a blend was the reduction in material costs when compared to a pure wool fabric [43]. However, there were also benefits to material properties when using Ardil as a blend, which *Silk & Rayon Magazine* reported on in 1945 [37]. Ardil was pitched to customers as a complimentary fibre that enhanced the properties of the other fibres it was blended with; it could be used to make lighter woollen fabrics, it was not subject to moth damage, and it was crease-resistant. Another key consideration in the production process highlighted by the *Silk & Rayon* article is that the variable length of fibres meant Ardil could be spun with existing machinery and, therefore, easily absorbed into company production lines [37]. Ardil was also found to be beneficial in hat making as, although it did not felt itself, it enhanced the felting process when used with wool [44].

In order to achieve the most from their new fibre, ICI encouraged industry partners to experiment with it. As a result, a wide array of Ardil blends were created, tested, and accessed [18]. It was suggested that blends with a higher percentage of Ardil would be better suited to dress fabrics, where washability was not as intensive or as frequent compared to shirting materials [43]. Ardil was blended in equal parts with wool, and just before the outbreak of war, sufficient fibre was produced on a laboratory scale to enable a number of suits to be made; a number of these suits were still being worn in the early 1950s [42]. Although *ICI Magazine* reported positively about the longevity of these early garments, anecdotal evidence suggests otherwise [45].

3.2.5. Ardil Factory

The ICI board approved Ardil for mass production in 1947. A GBP 2.1million budget for the project was agreed, and work began on the Ardil plant at the Dungan's site in Dumfries in 1949 [30]. The centrepiece of this site design was the Ardil Tower—a monolith to ICI's aspirations for their fibre (Figure 2b). The factory utilised a vertical assembly line, where all parts of the yarn-making process were conducted under one roof. The peanut protein meal was also stored onsite for ease of production. ICI were immensely proud of their new Nobel Division factory buildings, and boasted about the external "buff coloured" bricks and the "special acid-resisting floor" in an issue of *ICI Magazine* from 1950 [46]. Internally, the plant was fully tiled, as cleanliness was essential to the chemical processes of making Ardil. The description of the factory highlights the advancements made in contemporary RPF science, where the involvement of acid in the process would be highly controlled and minimised. When the plant was in full production, the output was around 9000 tonnes per year [42]. The site was ready for commercial production in 1951.

It took over 15 years from the initial idea developed by Astbury in his lab at the University of Leeds to the opening of Ardil's own dedicated plant. Throughout this time, numerous trials and testing of the material were carried out. Recommendations to blend Ardil were introduced to address flaws in its material strength, but despite this ICI still felt confident in the importance of the fibre they had invented.

3.3. Marketing and Public Reception of Ardil

3.3.1. General Public

Though ICI and textile manufacturers had great confidence in their product, they now faced a new challenge—marketing a new and unknown product to the public. After the Second World War, the public were keen to forget the hardship and rationing imposed on them, and to celebrate their post-war wealth. Although rationing of clothes continued until 1949, a desire for newness and convenience underpinned the marketing of many products, and the textile sector was no exception. Advances in science that had resulted from necessity during the war were now being applied to household products post-war. Textile journals from this period are filled with adverts for different coatings and finishes that ease the process of laundering [47]. The public were hungry for new materials and products that science could offer them, and the ease that they brought to their domestic lives. Ardil fit directly into this remit, with one advert declaring "Happy families—with 'Ardil'" (Figure 3a). Directly targeted at the housewife, this advert shows a domestic scene, and asserts that Ardil is a key ingredient to domestic bliss.

Though this was an era of new scientific ideas entering the domestic sphere, a textile made from peanuts was not easily accepted by the public. Many magazines and journals could not resist the humorous connotations of a fibre made from such an unlikely source. A headline from 1944 jubilantly declared that ICI had launched a "Monkey nut" fibre [34]. Another article reports that a wearer of an Ardil frock described it as "just nuts" [41]. Astbury was known to wear an overcoat made from Ardil during his lectures at the University of Leeds [48]; a cartoon published in the *Yorkshire Evening Post* in the same period [49] depicted the coat being pecked at by birds, further evidencing the humorous reception that this novel fibre would have had from the public (Figure 3b). It is notable that Astbury referred to the coat himself as made from "Monkey nuts", as he was not permitted to discuss ICI's new fibre by its brand name [32].

Figure 3. (**a**) Happy families—with "Ardil" advert circa 1955. (**b**) Cartoon of Astbury's "Monkey nut overcoat", *Yorkshire Evening Post*, 1944.

In fact, ICI were so concerned by these sardonic reviews from the press that they sold several garments made from Ardil to the public in secret [48]. Fearing that consumers would be put off by the origins of the yarn, they retailed several Ardil garments labelled as pure wool. This was part of wide-scale marketing test to see if the general public could notice the difference in the Ardil products.

3.3.2. The British Industries Fair

As well as advertising directly to consumers, ICI knew it was important to target the manufacturers who would use Ardil in their products. Established in 1915 by the Board of Trade, the British Industries Fair (BIF) was an esteemed showcase for British products and commerce. In 1953, ICI took a stand at the Earl's Court site of the BIF. ICI regularly showed their products at the BIF, but this was the first time they had shown a textile product, hoping to market it to manufacturers and the public alike. Stand No. R421/518 at the Earl's Court branch of the fair was an extremely large dual-aspect stand at the very centre of the exhibition [50]. ICI occupied more space than any other single exhibitor; the stand was designed by Hulme Chadwick, and was intended to show the fibre's "great versatility and exceptional properties" [51].

An advert for the upcoming show proudly declared Ardil to be "a new fibre of great importance to the textile industry" [52] (Figure 4a). The advert also hinted at several of the factors afflicting Ardil in its development so far. Ardil was referred to as a "new fibre" even though, as previously discussed, the development of Ardil had spanned nearly two decades already. Ardil was also described as "now available in bulk", alluding to the production and scaling problems faced at the plant. The BIF was so prestigious and the stand so intriguing that it was even visited by the new Queen, who was crowned the following month (Figure 4b) [53]. Tellingly, the Queen is reported to have asked "You tell us of all these remarkable properties. Surely there must be some snags?" [54].

(a) (b)

Figure 4. (a) "Ardil a new fibre of great importance to the textile industry", 1953; Catalyst Science Discovery Centre and Museum. (b) "Her Majesty is seen here examining 'Ardil' protein fibre" at the British Industries Fair, 1953; Catalyst Science Discovery Centre and Museum.

3.3.3. Designers Working with Ardil

In order to showcase Ardil at its best, ICI had commissioned some of the UK's leading textile designers to create fabrics using Ardil blends [51]; the pioneering woven furnishings designer Tibor Reich was one of them. It is perhaps no coincidence that Reich was a graduate of The University of Leeds, where the method for making Ardil was invented. The University's reputation in industry was strong, and Tibor's link to the institution may well have led ICI to commission him for their stand. Tibor Ltd. produced several fabrics using Ardil—most notably a design called History of Shapes (Figure 5), described by one reviewer as a "A tour de force woven on the Jacquard loom in [A]rdil, spun silk, and a metallic thread, which was then screen-printed with a narrative pattern" [55]. Today, one of the same pieces that hung on the BIF stand can be seen in the Tibor Ltd. private archive. The piece is regularly described as being made of Ardil, silk, and Lurex. Upon closer examination of the piece, it appears that Ardil is spun with silk to produce a unique yarn. The Victoria and Albert Museum also holds History of Shapes in their collection, although it is not identified as being made from Ardil. As a result of this research, the catalogue at the V&A has been corrected to record this piece as being made from Ardil.

Figure 5. Furnishing fabric History of Shapes, designed by Tibor Reich, Hungarian, Stratford-upon-Avon, circa 1953. © Victoria and Albert Museum, London.

As a designer, Reich was always interested in working with new materials and concepts. He was already using another novel material called Lurex—a yarn made from thin aluminium foil—in many of his textiles. After his experience using Ardil in History of Shapes, he continued to work with ICI, producing several commercial ranges of fabrics using Ardil (Figure 6a). Tibor's typical yarn palette would include wool yarns in different specifications, combined with viscose and Lurex wefts woven on a cotton warp. Ardil blended well with all of these fibres, and would have been complimentary to his existing products. Designs produced by Tibor Ltd. in Ardil included the Jacquards *Movemento*, *Granite*, and *Gazelle*, as well as power-woven designs, including Ardil Prince, Henley, and one of Tibor's most popular designs, California. Tibor was so taken with Ardil that he created a distinctive design celebrating peanuts, entitled Harvest (Figure 6b). Using the unusual technique that he had developed with History of Shapes, Harvest has a woven Jacquard background, which it was screen-printed onto. The motifs in the design are based on the process of harvesting peanuts. Tibor produced a similar design, called "Aluminium story", depicting the process of smelting aluminium [56].

Many other important names from 20th century textile design were keen to develop products using Ardil fibres. During the course of this research, it was discovered that one of the most notable manufacturers of 20th century textiles, Warner & Sons, also worked with Ardil fibre. Warner & Sons' power-woven record books show an Ardil fabric being produced in 1953 (Figure 7). The piece described in the log was produced for ICI in February 1953, suggesting that the intention was to showcase it on the Ardil stand at the BIF. The fabric was described as being green in colour and made form a blend of Ardil and silk. Further research is required to determine whether the Warner Textile Archive still holds any examples of this Ardil fabric. Other noted designers who lent their signature designs to Ardil fabrics included Jaqueline Groag, John Piper, Lucienne Day, and the sculptor Nicholas Vergette [57].

(a) (b)

Figure 6. (**a**) "Tibor weaves new 'Ardil' blend textures" advert circa 1954. (**b**) Furnishing fabric, Harvest, depicting the process of harvesting peanuts, designed by Tibor Reich, Stratford-upon-Avon, circa 1953.

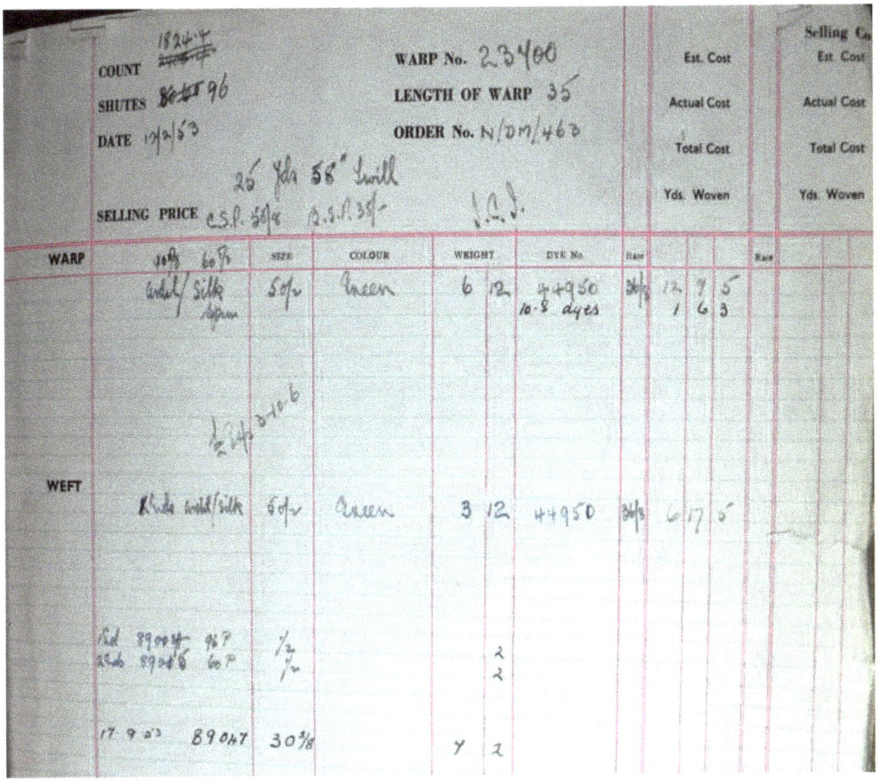

Figure 7. Power-woven fabric ledger, Warner & Sons, Warner Textile Archive, 1953. Reproduced with permission from the Warner Textile Archive, Braintree District Museum Trust.

3.4. Political and Economic Factors

Put simply, there are four main political and economic factors that governed the development of Ardil and can help to explain why it failed. These were the Second World War, the price of wool, the supply of peanuts, and the rise of petrochemical fibres. Whilst all of these are interlinked, each subject is tackled independently here to help understand the picture as a whole.

3.4.1. The Second World War

Although the Second World War was beneficial for development of many RPFs—most notably in Italy—the war actually postponed further development of Ardil, as much of ICI's technology was diverted towards the war effort. In fact, ICI's Drungans site in Dumfries, which would later become the Ardil plant, was originally created by ICI to produce munitions [58]. The peanut meal used to create Ardil was needed to supplement food supplies [39]. This significantly delayed the material developmental progress of Ardil. In contrast to the UK's postponement of Ardil, Italy doubled down on their output of RPFs during the war, clothing their army in Lanital uniforms. This may be in part due to the development of Ardil being several years behind that of Lanital; at the outbreak of war in 1939, Ardil was still only a lab creation. In 1944, development work on Ardil resumed, although ICI were still not able to produce Ardil in large quantities. "When the war ended, a small pilot plant was built and sufficient 'Ardil' produced to enable us to form some idea of its commercial possibilities" [42]. Slow development did not stop the press from eagerly reporting a 1944 press release from ICI. An article in *The Draper's Organiser* from January 1945 announced "New Wool-like Yarn from Nuts ... ICI's important contribution to synthetic fibres" [18].

3.4.2. The Wool Price

The war impacted the supply of raw materials, and there was a notable strain on wool supplies. At first, this was promising for Ardil and other RPFs, as it indicated that wool stocks could not be solely relied upon to keep up with demand for clothing [59]. Even the wool industry felt that scientific enhancement would likely play a part in the future of wool [60]. When wool prices began to climb to their most expensive in history at the end of the 1940s, it is likely ICI felt vindicated in their decision to invest to so heavily in Ardil. The cost of wool reached the highest it had ever been at "$125\frac{1}{2}$ d. per lb. on 26 January 1951" [61], roughly equivalent to GBP 41.80 per kg in today's money [62,63]. This steep elevation in price was caused by the government commandeering the British Wool Clip during the war. They continued to pay a set price for wool based on a pre-war price. This price did not reflect the cost of sheep farming and wool production at the time, so when the price cap was lifted at the end of the war the cost of wool rose significantly [64]. The demand for wool during the war period was also steadily increasing while the global output of wool fell, further inflating the price [65,66].

Data gathered compiled using Kreglinger and Fernau market reports shows the prices for wool at the London sales between 1924 and 1954 (Figure 8) [66], demonstrating the lack of recorded prices during the war period and subsequent elevation in price post-war. Unfortunately for ICI, this seemingly exponential rise in the price of wool did not last. The drop in wool prices globally [67] from 1951 onwards doubtless took a toll on Ardil, as industry partners were not as likely to try a more expensive, unfamiliar fibre over a reliable material such as wool. The commercial success of their product relied upon the price being highly competitive compared to wool and other fibres.

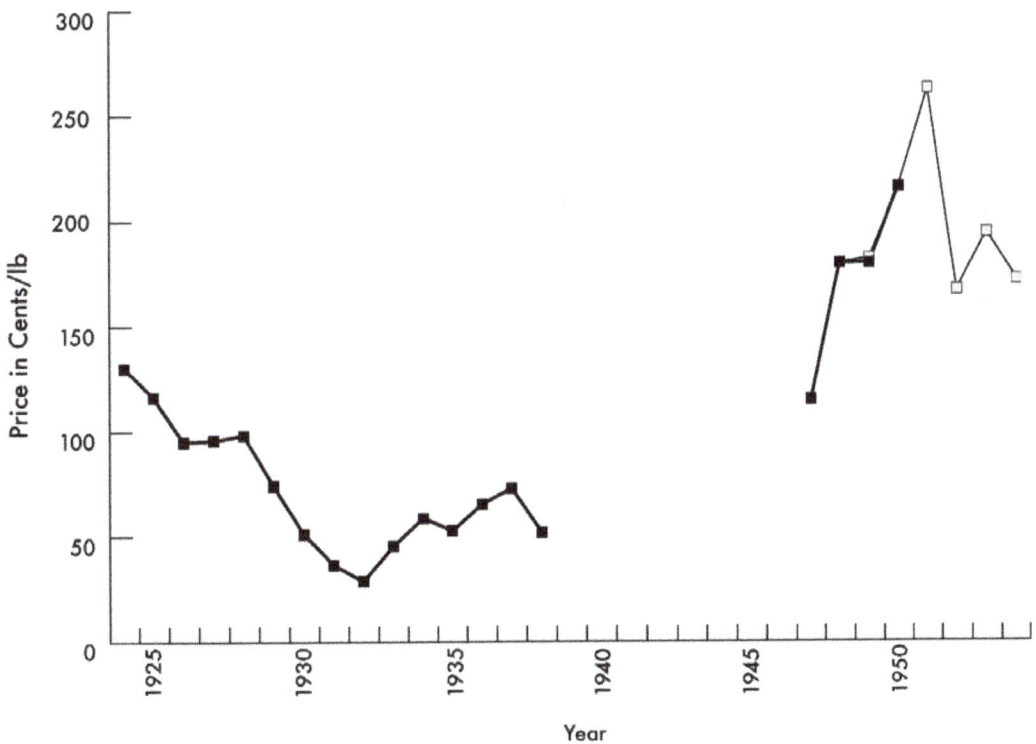

Figure 8. Graph showing the absence of a wool price during the war, and the initial spike and subsequent drop-off in wool prices post-war. Average price per pound and price differentials of fine wool at Boston and London markets, 1924–1954. N.B., Data taken from two separate sets, indicated by the colour change on the graph. Wool can be a difficult commodity to price, as it varies so much in quality, and is subject to fluctuations in yield. Data on the production and price of wool are therefore complicated to extrapolate, but the datasets used reflect the market trend as whole.

3.4.3. Supply of Peanuts

ICI had also not considered the difficulties they would face in finding the raw materials needed to scale up their production. At the same time as the peak in wool prices (July 1951), W Johnston at ICI wrote to Mr Greenhill of the Trades and Marketing Department of the Colonial Office, stating that he "would very much welcome any help or guidance which the Colonial office can give us in finding a source of consistent supply of good quality groundnuts ... to be used for the production of Ardil Protein Fibre" [68]. The problems ICI were experiencing came from the quality of protein yielded from the peanuts. They had already experimented with several varieties of peanut, each giving different results; the quality of the fibre produced was closely linked to the quality of the protein that could be extracted from the peanut meal [68]. As such, ICI found that they had to be highly selective about the peanut meal they used for the development of their product, forcing them to seek more reliable and consistent sources whilst narrowing their options. Johnston had contacted the Colonial Office off the back of a report about the available protein supplies for Ardil published by the Ministry of Food [69]; the report expressed concern about the growing demand for peanut meal from the textile industry, and warned that this demand had already come at the expense of "feeding-stuffs supplies" [69].

3.4.4. East Africa Groundnut Scheme

The failure of the East African Groundnut Scheme (EAGS) was also making headlines in 1951. Started in 1947, the EAGS was a proposed solution to post-war food shortages. The scheme entailed cultivating large quantities of peanuts in the shrubland of Tanganyika, now known as Tanzania [70]. It would be easy to assume a causal link between the demise of the EAGS and the demise of Ardil, but a direct link between the two is not clear [42]. The work on developing Ardil had begun long before the conception of the EAGS; concurrently, the EAGS was developed as a response to food shortages, not textile shortages. Correspondence from ICI shows that they were working with peanuts sourced from around the world, including India and China, as well as several other African countries in addition to Tanzania [70]. There is further correspondence showing an interest from ICI in the progress of the EAGS in early 1952 [71], but the demise of the EAGS was already in motion [72]. Instead, it would be more accurate to suggest that the simultaneous rise and fall of both Ardil and the EAGS speak to the wider trend of peanuts as a commodity at the time, and the promise that they held as both a fibre resource and a foodstuff.

3.4.5. The Rise of Petrochemicals

As referenced in Figure 1, the development of Ardil and other regenerated protein fibres coincided with the development of new fibres from the petrochemical industry. The synchronicity of these emerging fibre technologies is perfectly encapsulated by ICI itself; at the same time as they were working on Ardil, they were developing another patented textile fibre—Terylene. Terylene is the ICI brand name for polyethylene terephthalate (PET)—a fibre derived from petrochemicals. These two textile products from ICI were developed concurrently; the 1953 BIF also served as a launch pad for Terylene, with one side of the dual-aspect stand being used for each fibre. While the Ardil factory at Ardeer was being built, a pilot plant was producing Terylene in Lancashire. A full-scale production facility dedicated to Terylene was completed at the end of 1954. It was capable of producing up to 10,000 tonnes of fibre a year, and was situated in Wilton, in what is now Teesside [73]. ICI invested heavily in and widely promoted both products; however, Terylene ultimately won, succinctly summarised by this 1957 headline in *The Outfitter* "Terylene output to be doubled—but it's the end of Ardil" [74].

4. Discussion

Ardil represents a unique and important case study in the development of RPFs. Its lifespan only lasted 22 years from conception to demise, but during that time Ardil was hugely prevalent. From the first experiments carried out by Astbury in his lab at the University of Leeds to the practically overnight closure of the Ardil factory in 1957, Ardil's story is full of Hollywood-like twists. The huge amounts of finance and time invested show how strongly Ardil's backers believed that this could be the future of material science. At its best, Ardil was a fibre that could simplify manufacturing techniques, address material shortages, and provide consistent and superior textile qualities such as drape and lustre. At its worst, Ardil was an expensive experiment that produced inconsistent results, was not scalable, could never be price-competitive with equivalent fibres such as wool, and lacked material longevity.

Unfortunately, the popularity and success of petrochemical textiles meant that RPFs were largely abandoned, and much of the information and research into them became obscured. In the 21st century, we are seeing a huge reversal in the popularity of petrochemical fibres as we discover more about their detrimental impact to the planet. Had we known then what we know now, research into RPFs might not have died out in favour of petrochemical fibres. RPFs hold huge potential for the future of sustainable textiles. A vast quantity of research into this field already exists, but is currently inaccessible for researchers wishing to build upon it. Archival research has provided more detail and clarity as to the reasons why Ardil failed, uncovering previously unseen or little-known material that has helped to build a clearer picture of the history of this once-celebrated fibre.

This methodology of reflecting on lessons from societies past has been mentioned by the Centre for Circular Design, Chelsea College of Arts, as part of their TED's TEN toolkit for sustainable design [75].

Ardil is a prime example of how and why RPFs met their untimely end as a result of myriad factors. Perhaps the most important of these was the poor performance of these fibres in comparison to others available on the market. It was clear that manufacturers were not always as keen to work with these new manmade fibres as the companies who promoted them. Another 1957 article in *The Outfitter* entitled "Test fibres for longer before we sell them" described how manmade fibres had "bedevilled the outfitting trade since the war" [76]. Customers were also not as keen to buy Ardil as ICI had hoped, with sales failing to grow. The struggle for new and experimental fibres to be accepted into a wider market is an important lesson to be taken forward into contemporary RPF development; the inherent poor fibre strength that plagued manufacturers at the time could now be seen as an opportunity, aligning RPFs with alternative, faster fashion cycles [77].

The issues Ardil faced with the supply of the peanut raw material are also relevant to contemporary RPF research. After the end of the hostilities in WW2 and the subsequent dissolving of the British Empire, the volume of peanuts imported to the UK reduced massively, meaning that the economics of utilising peanut waste for textile manufacturing were no longer favourable. While it was also hoped that Ardil would be cheap to manufacture, as it largely utilised byproducts from the food industry, it became clear that ICI needed to be more selective with the proteins they used in order to produce the best-quality fibre. Therefore, they began to move towards sourcing peanuts directly for producing Ardil, which was far less cost-effective than using peanut waste as they had initially planned. This highlights the need for robust future-proofing when designing a circular economy, ensuring that feedstocks are ideally locally sourced as well as actual waste streams.

Looking into the future of incorporating regenerated protein fibres into a circular economy, the environmental impact of the production process in terms of both the chemicals used and the volume of water consumed has to be considered; there has been a lot of contemporary research done into the replacement of formaldehyde as a crosslinking agent with more sustainable options, such as polycarboxylic acids; and methods for regenerating the water used within the process to try and "close the loop" could help alleviate issues with excessive water consumption. More research would have to be performed to determine the full environmental impact of these fibres through LCA, to identify which areas need to be improved and how they could be improved from an environmental perspective.

5. Conclusions

From a modern sustainability perspective, the prospect of importing vast quantities of produce from overseas is also problematic, with the correspondingly large environmental impact of transportation of the goods. It is much more attractive to look at processing the feedstock at a local level and utilising the waste as close to the production and processing as possible. Within the UK, there is no longer any potential for utilisation of large quantities of peanut waste, but the UK does have other forms of protein waste that could potentially be utilised. The majority of food consumed within the UK is actually dairy, accounting for roughly 27% of the food eaten [78]; therefore, a correspondingly large volume of waste is generated, with over 330,000 tonnes of milk being wasted every year, giving potential for RPFs from milk—such as casein fibres—to be explored. However, it should be noted that the majority of the milk being wasted in the UK is being generated in homes (90%), which would give rise to logistical problems for collection and ensuring that a uniform feedstock is obtained [79]. Casein-based RPFs were pioneered within Europe at roughly the same time as Ardil, with fibres such as Lanital having a similarly short-lived lifetime during and after WW2, although more contemporary research has been conducted into casein-based fibres [77].

While the demise of the UK-based RPF Ardil was marked by the sudden lack of peanuts being exported into England, this does not represent a reduction in the peanut waste being generated globally. Indeed, peanut production has seen a marked increase in recent years, due to an increase in the popularity of peanut oil. Global peanut production has increased from 31.4 Mt in 2000 to 48.1 Mt in the 2019/2020 season, representing a > 50% increase in the last 20 years [80]. Global peanut oil production has increased from 4.5 Mt in 2000 to 6.5 Mt in 2021/2022, representing a roughly 43% increase [81]. When processing peanuts for oil, the waste in the form of peanut meal can be as high as 70%; the global production of peanut meal in 2019–2020 was 7.7 Mt. As discussed previously, this peanut meal has a 53.3% average protein content, representing a raw waste protein mass of roughly 4.1 Mt; currently, the main use of this feedstock is animal feed. However, this is a low-value valorisation route for this waste feedstock, and also poses potential issues with contamination with aflatoxins. Peanuts are particularly susceptible to contamination by *Aspergillus flavus* and *Aspergillus parasiticus* fungi, which produce aflatoxins that in high enough doses are lethal to both humans and animals, and low doses can still cause myriad diseases, including cancer in humans, and have been shown to reduce weight gain and milk and egg production, as well as causing contamination of milk in animals. These issues do not stop peanut meal being a highly effective animal feed component, but they do highlight that it is not a perfect solution to this waste stream, and there is historical precedent to allow for this huge waste stream to help alleviate the modern world's reliance on non-renewable textiles [82].

For both of the feedstocks discussed, as well any waste utilised with the intention of replacing conventional fibres in the textile industry, care must be taken regarding the volumes of feedstock required. The textile industry is huge, and if regenerated protein fibres were to be accepted as a replacement for non-renewable fibres, the feedstock would need to be able to keep up with demand. This issue would require a collaborative effort across multiple disciplines to determine whether replacement with RPFs would be feasible, with potential future work being focused on the opportunities and drawbacks for theses fibres through SWOT analysis and LCA. It would also be important to learn from past mistakes and use the critical analysis of why Ardil failed during its initial conception, and what factors could be used to avoid such failure in the future.

Author Contributions: Conceptualisation, H.A.G., J.A.H., M.S., R.S.B. and V.K.; methodology, H.A.G., J.A.H. and M.S.; validation, H.A.G., J.A.H., M.S., R.S.B. and V.K.; formal analysis, H.A.G. and J.A.H.; investigation, H.A.G. and J.A.H.; resources, R.S.B. and V.K.; writing—original draft preparation, H.A.G. and J.A.H.; writing—review and editing, R.S.B., M.S. and V.K.; visualisation, H.A.G. and J.A.H.; supervision, R.S.B. and V.K.; project administration, R.S.B. and V.K.; funding acquisition, R.S.B. and V.K. All authors have read and agreed to the published version of the manuscript.

Funding: This research was funded by the Business of Fashion, Textiles, and Technology (www.bftt.org.uk) (accessed on 16 February 2022), a Creative R&D Partnership funded by the Creative Industries Clusters Programme funded by the Industrial Strategy, and delivered by the Arts and Humanities Research Council on behalf of UK Research and Innovation, under grant no. AH/S002804/1.

Institutional Review Board Statement: Not applicable.

Informed Consent Statement: Not applicable.

Acknowledgments: We would like to thank the Catalyst Science Discovery Centre and Museum; Tibor Ltd.; the Victoria and Albert Museum; and the Warner Textile Archive, Braintree District Museum Trust for their help with this research and the use of their images.

Conflicts of Interest: The authors declare no conflict of interest.

References

1. Committee H of CEA. Fixing Fashion: Clothing Consumption and Sustainability. House of Commons: European Union News. 2019. Available online: https://publications.parliament.uk/pa/cm201719/cmselect/cmenvaud/1952/report-summary.html (accessed on 9 February 2022).
2. Preferred Fiber & Materials Market Report. 2021. Available online: https://textileexchange.org/preferred-fiber-and-materials-market-report/ (accessed on 14 March 2022).

3. Matharu, A.; Melo, E.; Houghton, J.A. Green chemistry: Opportunities, waste and food supply chains. In *Routledge Handbook of the Resource Nexus*; Routledge: Abingdon on Thames, UK, 2017; pp. 457–467.
4. Remy, N.; Speelman, E.; Swartz, S. *Style That's Sustainable: A New Fast-Fashion Formula*; McKinsey Global Institute: Seoul, Korea, 2016. Available online: https://search.proquest.com/docview/2372072480 (accessed on 9 February 2022).
5. Bhardwaj, V.; Fairhurst, A. Fast fashion: Response to changes in the fashion industry. *Int. Rev. Retail. Distrib. Consum. Res.* **2010**, *20*, 165–173. [CrossRef]
6. Bhardwaj, V.; Manchiraju, S. The Role of Impulse Buying, Hedonism, and Consumer Knowledge Towards Sustainable Consumption of Fast Fashion. 2017. Available online: https://dr.lib.iastate.edu/handle/20.500.12876/51826 (accessed on 9 February 2022).
7. Moazzem, S.; Daver, F.; Crossin, E.; Wang, L. Assessing environmental impact of textile supply chain using life cycle assessment methodology. *J. Text. Inst.* **2018**, *109*, 1574–1585. [CrossRef]
8. Valuing Our Clothes: The Cost of UK Fashion. WRAP, London. 2017. Available online: https://wrap.org.uk/sites/default/files/2020-10/WRAP-valuing-our-clothes-the-cost-of-uk-fashion_WRAP.pdf (accessed on 9 February 2022).
9. Britten, F. Fashion Brands Pause Use of Sustainability Index Tool over Greenwashing Claims. The Guardian. Available online: https://www.theguardian.com/fashion/2022/jun/28/fashion-brands-pause-use-of-sustainability-index-tool-over-greenwashing-claims (accessed on 4 July 2022).
10. Cotton Industries. WWF. Available online: https://www.worldwildlife.org/industries/cotton (accessed on 16 March 2022).
11. Ferrigno, S.; Guadagnini, R.; Tyrell, K. Is Cotton Conquering Its Chemical Addiction? Available online: https://www.pan-uk.org/cottons-chemical-addiction/ (accessed on 16 March 2022).
12. Shepherd, H. *Thirsty for Fashion?* Soil Association: Bristol, UK, 2019.
13. Briefing On Silk. Common Objective. Available online: https://www.commonobjective.co/article/fibre-briefing-silk (accessed on 16 March 2022).
14. Fishwick, M. *A Carbon Footprint for UK Clothing and Opportunities for Savings*; WRAP: Banbury, UK, 2012.
15. Bat-Erdene, B. Desertification in Mongolia: Untangling Problems, Effects, and Policy Solutions. 2019. Available online: https://repository.arizona.edu/browse?value=Bat-Erdene%2C+Bat-Orgil&type=author (accessed on 16 March 2022).
16. Middleton, N. Rangeland management and climate hazards in drylands: Dust storms, desertification and the overgrazing debate. *Nat. Hazards* **2018**, *92*, 57–70. [CrossRef]
17. Traill, D. Clothing an Expanding Population. *J. Text. Inst.* **1951**, *42*, 221–236. [CrossRef]
18. The Drapers' Organiser. New Wool-like Yarn from Nuts. *The Drapers' Organiser*, January 1945; pp. 22–23.
19. Swales, J.N. Protein and synthetic fabric fibres. *Furnishing World*, 25 November 1955; p. 47.
20. Wormell, R.L. *New Fibres From Proteins*; Academic Press: Cambridge, MA, USA, 1954.
21. Stenton, M.; Houghton, J.A.; Kapsali, V.; Blackburn, R.S. The Potential for Regenerated Protein Fibres within a Circular Economy: Lessons from the Past Can Inform Sustainable Innovation in the Textiles Industry. *Sustainability* **2021**, *13*, 2328. Available online: https://www.mdpi.com/2071-1050/13/4/2328 (accessed on 9 February 2022). [CrossRef]
22. O'Brian, R. Wartime Textile Adjustments. *J. Home Econ.* **1942**, *34*, 514.
23. Heard, E. Wartime Developments in Textiles and Clothing. *J. Home Econ.* **1942**, *34*, 427–432.
24. Brooks, M.M. *Ardil: The Disappearing Fibre? Saving the 20th Century*; Canadian Conservation Institute: Ottawa, ON, Canada, 1993; pp. 81–93.
25. Maxwell, A. *Patriots Against Fashion*; Springer: Berlin/Heidelberg, Germany, 2014.
26. Imperial Chemical Industries. *The Dyeing of "Ardil" Protein Fibres and Ardil Fibre Unions. Imperial Chemical Industries*; Imperial Chemical Industries: London, UK, 1957.
27. Ernest, R.; Kaswell, E.R. *Textile Fibers, Yarns and Fabrics*; Reinhold Publishing Corporation: New York, NY, USA, 1953.
28. Reader, W.J. *Imperial Chemical Industries: A History: Volume 1: The Forerunners, 1870–1926*, 1st ed.; Oxford University Press: Oxford, UK, 1970; Volume 1.
29. International Textiles. A visit to Leeds' Textile University. *Int. Text.* **1944**, *6*, 30–31.
30. Reader, W.J. *Imperial Chemical Industries: A History*; Oxford University Press: Oxford, UK, 1975; Volume 2.
31. Chibnall, A.; Bailey, K.; Astbury, W. Improvements in or Relating to the Production of Artificial Filaments, Threads, Films, and the Like. Great Britain Patent GB29161/35A, 22 October 1935.
32. Kersten, T.; Hall, K.T. *The Man in the Monkeynut Coat*; Oxford University Press: Oxford, UK, 2014.
33. ICI Magazine. Our Contributors. *ICI Mag.* **1951**, *29*, 173.
34. International Textiles. ICI launch "Monkeynut" Fibre. *Int. Text.* **1944**, *12*, 44.
35. Traill, D. Manufacture of Filaments from Vegetable Globulin. United. States Patent US2358427A, 4 August 1941.
36. Traill, D. Improvements in or Relating to Films, Filaments and Other Shaped Articles Made by Hardening Proteins Coagulated from Their Solutions. Great Britain Patent GB8725/37A, 24 March 1937.
37. Silk & Rayon. Ardil: ICI Fibre from Ground Nut Protein. *Silk & Rayon*, January 1945; pp. 31–99.
38. *Annual Report on the Social and Economic Progress of the People of the Gambia 1935. Colonial Reports Annual*; His Majesty's Stationary Office: London, UK, 1935.
39. World Fiber Survey. His Majesty's Stationary Office; London 1947. H.M. Stationery Office. 1947. Available online: https://books.google.co.uk/books?id=9ImSzgEACAAJ (accessed on 9 February 2022).
40. *Report of the Administrator of Agricultural Research. United States*; Agricultural Research Administration: Beltsville, MD, USA, 1943.

41. The Drapers' Organiser. Peanut Dress? *The Drapers' Organiser*, June 1944; p. 27.
42. ICI Magazine. From Groundnuts to Ardil. *ICI Mag.* **1951**, *29*, 174.
43. Skinner's Silk and Rayon Record. Ardil Fibre/Viscose Rayon Staple/Nylon Staple Blend Fabrics. *Skinner's Silk and Rayon Record*, December 1956.
44. Barr, T.; Haigh, D. Wool protein fibre blends in felt manufacure. *J. Text. Inst. Proc.* **1952**, *43*, 593–603. [CrossRef]
45. Byrne, M. The Story of Scotland's Own Fibre. Available online: https://www.icidumfriesphotos.com/home (accessed on 4 November 2021).
46. Braham, J.E. The New Ardil Factory. *ICI Mag.* **1950**, *28*, 250–251.
47. The Outfitter. A cloth finish to counter body odours. *The Outfitter*, 28 September 1957; Volume 7.
48. Sunday Dispatch Reporter. Hundreds of Women Wear Monkey Nut Suits—But Do Not Know It. *Sunday Dispatch Reporter*, February 1952; 3.
49. British Newspaper Archive. Yorkshire Evening Post. *British Newspaper Archive*, 18 January 1944; p. 5.
50. Board of Trade. 1953 British Industries Fair Catalogue. *Board of Trade*, April 1953.
51. ICI Magazine. I.C.I. at the B.I.F. *ICI Mag.* **1953**, *31*, 151–152.
52. Board of Trade. Ardil Advert. 1953 British Industries Fair Catalogue. *Board of Trade*, April 1953.
53. ICI Magazine. The Queen. *ICI Mag.* **1953**, *31*, 181–182.
54. ICI Magazine. Rayon to Terylene. *ICI Mag.* **1953**, *31*, 226–229.
55. *Untraceable Find. Newspaper Clippings, Tibor Ltd. Archive*; Balding & Mansell, Wisbech: London, UK, 1950.
56. Ellis-Petersen, H. Him Indoors: Tibor Reich, the Designer Who Brightened up Britain Forever. The Guardian. 2016. Available online: https://www.theguardian.com/artanddesign/2016/feb/01/tibor-reich-the-whitworth-manchester-interiors-britain-designer-bright-colour (accessed on 2 February 2022).
57. International Textiles. Ardil. *International Textiles*, April 1953; pp. 75–77.
58. Pollock, I. *Factory Celebrates 70 Years*; Dumfries Standard Weekly: Glasgow, UK, 2009.
59. Dillon, J.H.; Brown, H.F. An Introduction to the Preparation and Properties of Yarns Containing Ardil Protein Fibre. *J. Text. Inst. Proc.* **1952**, *43*, 584–592. [CrossRef]
60. McCann, C. *Wool: No Easy Optimism*; International Textiles: Beverly Hills, CA, USA, 1944.
61. Wool Prices 1951. Hansard, UK Parliament. Available online: https://hansard.parliament.uk/Commons/1951-02-15/debates/593ec0fa-8b65-4a7f-8593-5efc4c3db0a4/WoolPrices (accessed on 26 January 2022).
62. Historical UK Inflation Rates and Price Conversion Calculator. Available online: https://iamkate.com/data/uk-inflation/ (accessed on 27 January 2022).
63. Currency Converter: 1270–2017. Available online: https://www.nationalarchives.gov.uk/currency-converter/#currency-result (accessed on 27 January 2022).
64. Wool Prices 1940. Hansard, UK Parliament. Available online: https://hansard.parliament.uk/Lords/1940-08-01/debates/fc036680-1fec-4106-a0bb-9f1828ba2790/WoolPrices (accessed on 9 February 2022).
65. Blau, G. Wool in the World Economy. *J. Text. Inst. Proc.* **1946**, *37*, 454–456. [CrossRef]
66. Hermie, A.M. *Prices of Apparel Wool Prices of Apparel Wool*; Technical Bulletin No. 1041; Bureau of Agricultural Economics: West Lafayette, IN, USA, 1951.
67. Nattrass, N.; Conradie, B. Jackal Narratives: Predator Control and Contested Ecologies in the Karoo, South Africa. *J. S. Afr. Studies* **2015**, *4*, 41. [CrossRef]
68. Johnston, W. *Letter to M.A. Greenhill. CO 852/1159/8, Supplies for "Ardil Protein Fibre"*; The National Archives: Richmond, UK, 1951.
69. Animal Foodstuffs Division of the Ministry of Food. *N.R. (51) 19 Protein Supplies for Ardil*; The National Archives: Richmond, UK, 1951.
70. Wood, A. The Groundnut Affair. Bodley Head. 1950. Available online: https://books.google.co.uk/books?id=dvg9AAAAYAAJ (accessed on 9 February 2022).
71. Greenhill, M.A. *Letter to W. Johnston. CO 852/1159/9*; The National Archives: Richmond, UK, 1952.
72. Westcott, N. *Imperialism and Development: The East Africa Groundnut Scheme and its Legacy*; Woodbridge: Suffolk, UK, 2020.
73. The Times Survey of the British Industries Fair. The Advance of Cotton. *The Times Survey of the British Industries Fair*, 14–17 May 1954.
74. The Outfitter. Terylene output to be doubled—but it's the end of Ardil. *The Outfitter*, 21 September 1957; p. 7.
75. The Ten. Centre for Circular Design. Available online: https://www.circulardesign.org.uk/research/ten/ (accessed on 4 July 2022).
76. The Outfitter. Test fibres for longer before we sell them. *The Outfitter*, 5 October 1957.
77. Stenton, M.; Kapsali, V.; Blackburn, R.S.; Houghton, J.A. From Clothing Rations to Fast Fashion: Utilising Regenerated Protein Fibres to Alleviate Pressures on Mass Production. *Energies* **2021**, *14*, 5654. [CrossRef]
78. Jeswani, H.K.; Figueroa-Torres, G.; Azapagic, A. The extent of food waste generation in the UK and its environmental impacts. *Sustain. Prod. Consum.* **2021**, *26*, 532–547. [CrossRef]
79. Opportunities to Reduce Waste along the Journey of Milk, from Dairy to Home | WRAP. Available online: https://wrap.org.uk/resources/case-study/opportunities-reduce-waste-along-journey-milk-dairy-home (accessed on 16 February 2022).
80. USDA. World Agricultural Production. February 2022. Available online: https://www.fas.usda.gov/data/world-agricultural-production (accessed on 4 February 2022).

81. Oilseeds: World Markets and Trade. February 2022. Available online: https://apps.fas.usda.gov/psdonline/circulars/oilseeds.pdf (accessed on 14 February 2022).
82. Heuzé, G.; Tran, D.; Bastianelli, F.; Lebas, H.T. Peanut Meal. Feedipedia—Animal Feed Resources Information System. 2018. Available online: https://www.feedipedia.org/node/699www.feedipedia.org/node/699 (accessed on 9 February 2022).

Article

Improving Mechanical Textile Recycling by Lubricant Pre-Treatment to Mitigate Length Loss of Fibers

Katarina Lindström *, Therese Sjöblom, Anders Persson and Nawar Kadi

Department of Textile Technology, Faculty of Textiles, Engineering and Business, University of Borås, SE-501 90 Borås, Sweden; Therese.sjoblom@gmail.com (T.S.); anders.persson@hb.se (A.P.); nawar.kadi@hb.se (N.K.)
* Correspondence: katarina.lindstrom@hb.se

Received: 14 August 2020; Accepted: 15 October 2020; Published: 20 October 2020

Abstract: Although there has been some research on how to use short fibers from mechanically recycled textiles, little is known about how to preserve the length of recycled fibers, and thus maintain their properties. The aim of this study is to investigate whether a pre-treatment with lubricant could mitigate fiber length reduction from tearing. This could facilitate the spinning of a 100% recycled yarn. Additionally, this study set out to develop a new test method to assess the effect of lubricant loading. Inter-fiber cohesion was measured in a tensile tester on carded fiber webs. We used polyethylene glycol (PEG) 4000 aqueous solution as a lubricant to treat fibers and woven fabrics of cotton, polyester (PES), and cotton/polyester. Measurements of fiber length and percentage of unopened material showed the harshness and efficiency of the tearing process. Treatment with PEG 4000 decreased inter-fiber cohesion, reduced fiber length loss, and facilitated a more efficient tearing process, especially for PES. The study showed that treating fabric with PEG enabled rotor spinning of 100% recycled fibers. The inter-fiber cohesion test method suggested appropriate lubricant loadings, which were shown to mitigate tearing harshness and facilitate fabric disintegration in recycling.

Keywords: textile recycling; yarn spinning; inter-fiber cohesion; lubricant; mechanical tearing

1. Introduction

Materials used in textiles have a negative environmental impact; two thirds originate from petrochemicals, while the production and processing of cotton (CO) requires a high amount of water and generates a lot of wastewater [1]. The high demand for textiles result in increased textile waste generation [2]. By using the textile waste as an asset and re-using it as a raw material in textile production, the environmental impact of the textile industry could be reduced. By mechanical tearing, recovered fibers can be re-assembled into yarn and, subsequently, into textiles. However, fiber spinnability and yarn strength is largely affected by the fiber length, which decreases during tearing [3]. Aronsson and Persson [3] investigated how the condition of worn garments affected the quality of the recycled cotton fibers from post-consumer denim and single jersey fabrics. They found that more heavily worn garments had shorter fiber lengths than the less worn ones. However, after tearing, the fiber length difference between the different degrees of wear was insignificant for the denim while the single jersey actually recorded significantly shorter fibers for the less worn fabrics. The authors' reverse engineering approach revealed that the single jersey was knitted from fine ring-spun yarns whereas the denim consisted mainly of rotor spun yarns. They argue that the yarn construction is more important than the degree of wear for the tearing outcome [3].

It has been reported that the decrease in fiber quality during textile tearing makes it necessary to blend the recycled fibers with virgin fibers to enable the spinning of yarn [4–8]. Further, the choice of yarn spinning method is limited by the short fiber content. Ring spinning cannot handle such short fibers. Thus, rotor or friction spinning is used. Limited research has focused on the spinning of fibers

Sustainability **2020**, *12*, 8706; doi:10.3390/su12208706   www.mdpi.com/journal/sustainability

from recycled textiles. Merati and Okamura [4] used friction spinning to blend recycled fibers with 51% virgin CO, adding a PES filament core that increased the regularity and strength in a yarn of count 30 tex. Wanassi et al. [5] used rotor spinning to produce yarn of 50/50 recycled and virgin CO fibers. Pre-consumer waste from, e.g., the CO spinning industry have some similarities to torn fibers; the main one being a high amount of short fibers.

Mohamed Taher et al. [6] managed to incorporate 25% CO waste into rotor spun yarn without any change in appearance, regularity, or uniformity. Duru and Babaarslan [7] investigated the effect of the opening roller speed on a 60/40 blend of virgin PES and waste fibers. Khan, Hossain, and Sarker [8] found that the blend ratio significantly affected the yarn strength of rotor spun yarn, although higher cylinder speed could increase the quality with a higher percentage of added waste. Several researchers have performed similar work and adopted similar methods to spin yarn from recycled fibers. The yarn spinning process and final textile character is greatly influenced by the frictional behavior of the fibers [9,10].

The friction coefficient is not easily measured between staple fibers due to the difficulties of controlling a continual addition of normal load on the fibers. For comparison purposes, the frictional behavior of staple fibers can be measured with a cohesion test. Inter-fiber cohesion is often defined as the ability of a fiber arrangement to hold its shape or the energy needed to separate a fiber assembly [11,12]. Fiber cohesion is influenced by fiber friction, the shape of the fiber, and the flexural rigidity of the fibers, as well as the fiber length and denier [11,13]. As the inter-fiber cohesion is dependent on fibers' individual properties as well as the arrangement of the fibers, cohesion tests are typically comparative tests. Fiber cohesion of staple fibers can be measured dynamically by measuring the mean drafting force of slivers or rovings with, e.g., a Westpoint cohesion tester or a modified rotorring. Dynamic cohesion tests measure the force of straightening the fibers and, over a certain draft ratio, the sliding of fibers, but rarely reach the maximum force needed to break a fiber arrangement. A cohesion test was developed by Barella in 1953 that measured the minimum twist of cohesion, which is the minimum twist needed to hold a sliver or yarn together during the tensioning from a weight [14,15]. A static cohesion test can be performed with a tensile tester on a sliver, roving, or carded web [16–18]. The American standard ASTM D2612-99 [17] describes a static cohesion test on slivers and tops. Scardino and Lyons [18] performed a similar test on carded webs where they normalized the maximum force to the linear density and called it the *maximum cohesive tenacity*.

When performing fiber cohesion tests on fiber samples treated with different loadings of finishes and lubricants, the difference between samples shows the lubrication effect on inter-fiber friction [19]. An encouraging effect of lubricant on inter-fiber friction has been shown when the fibers were treated with an optimum lubricant concentration. The inter-fiber friction is at its lowest when the fibers are treated at this concentration, and the friction increases on further increase of lubricant after which the friction may decrease; this effect has been reported in previous studies [9,20,21]. The explanation for this phenomenon is that there are three zones for the loading of lubrication: (a) low concentration, where a mono-layer is formed; (b) intermediate zone, where hydrodynamic resistance is created; and (c) a high loading of lubricant with hydrodynamic flow conditions [9]. Depending on fiber type and lubricant, the optimum lubricant concentration is usually found between 0.1 and 0.5% [9,20,21].

Polyethylene glycol (PEG) is used in the textile industry to lower friction and, e.g., increase spinnability or to improve the hand of a fabric [22]. However, it is not a lubricant used in the industry with CO. Other lubricants often used are hydrophobic oils; however, these are attracted to synthetic fibers, which makes removal difficult [23]. PEG is non-toxic and soluble in water, which facilitate ease of application and removal [24,25].

Due to the harsh tearing process, which shortens the fiber length, mechanical recycling of textiles has traditionally resulted in low value products, such as rags or stuffing for insulation. To increase the quality of recycled fibers and enable reassembling to yarn and textiles, the fiber length needs to be retained to a larger degree. This paper describes how a comparative method to quantify inter-fiber cohesion was utilized to suggest suitable PEG 4000 pretreatment loading on off-the-shelf fabrics before mechanical recycling. The effects of variation of the PEG loading upon tearing outcome were studied for CO, PES, and cotton/polyester (CO/PES) plain weave fabrics.

2. Materials and Methods

2.1. Materials

For inter-fiber cohesion measurement, carded webs of CO and PES staple fibers were made. CO fibers were 25 mm mean length and had a linear density of 2.0 dtex, which was calculated from micronaire. PES fibers of 52 mm with a fiber linear density of 2.2 dtex were utilized, as per information given by supplier.

The fabrics used for tearing were 100% CO, 100% PES, and 50/50% CO and PES blend (CO/PES) of plain weaves supplied by Whaleys Bradford Ltd.; see Table 1 for analysis and reverse engineering details. The metric measurement of the weight of a fabric was measured after laundering, using a 1 dm^2 cutter.

Table 1. Fabric specification.

Material	Fabric Surface Weight	Warp	Weft	Fiber Length (mm)	
				Warp	Weft
CO	170 g/m^2	Ring spun, 29 tex 26 threads/cm	Ring spun, 28 tex 25.8 threads/cm	20.6	27.6
PES	152 g/m^2	Multifilament, 18 tex 29.4 threads/cm	Ring spun, 39 tex 22.2 threads/cm	Filament	28.0
CO/PES	142 g/m^2	Ring spun, 23 tex 30.7 threads/cm	Ring spun, 23 tex 25.4 threads/cm	25.4	28.8

The materials were treated with polyethylene glycol (PEG) 4000, supplied by Merck. The treatment solutions were prepared by mixing PEG with deionized water heated to 60 °C.

2.2. Methods

2.2.1. Fiber Web Treatment

Fiber batches of 40 g were opened in a La Roche edge opener and carded once in a Mesdan 337A laboratory carding machine. Each batch was treated with 20 mL PEG solution with concentrations in accordance with Table 2. The blend CO/PES constituted 50% of each fiber. PEG concentrations were based on fiber weight, e.g., for 0.25 wt.%, 100 mg PEG was diluted with deionized water. The solution was sprayed on the fibers using a high pressure spray gun set at a pressure of 0.02 MPa and subsequently left to dry for 2 h at room temperature. After drying, two 40 g fiber batches were carded twice into 80 g webs used for inter-fiber cohesion testing; see Figure 1 for a process overview.

Table 2. Concentrations of PEG 4000 in treatment solutions of fibers.

	CO Fiber	PES Fiber	CO/PES Fiber
	0.0	0.0	0.0
	0.25	0.25	0.5
PEG 4000	0.5	0.5	0.75
conc. wt.%	0.75	0.75	1.25
	1.0	1.0	
	1.25	1.25	

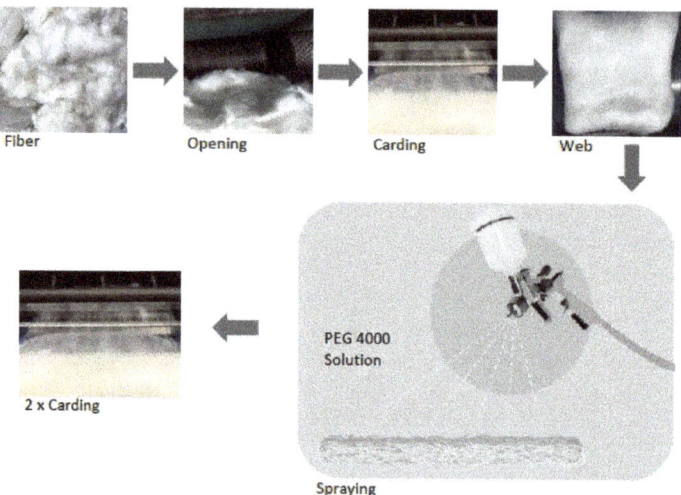

Figure 1. Treatment of fiber webs.

2.2.2. Fabrics Treatment

The concentrations for fabric treatments were chosen following the results from the inter-fiber cohesion tests; see Table 3. We used the higher percentage of PEG to verify the effect in the tearing process and to see if the fiber length would be preserved with the higher concentration.

Table 3. Concentrations of PEG in treatment solutions of fabrics.

	CO Fabric	PES Fabric	CO/PES Fabric
PEG conc. wt.%	0.0	0.0	0.0
	0.1	0.2	0.1
	0.3	0.7	0.5

All fabrics were laundered prior to treatment to get rid of any chemicals used in manufacturing, such as sizing. Laundering was performed according to standard *ISO 6330:2012 Textiles—Domestic washing and drying procedures for textile testing*. Household detergent was used. Washing programs from Annex B were followed, both with one washing cycle and four rinses. The program 6N was used for PES and CO/PES, with a washing temperature of 60 °C. Cotton was washed twice with a temperature of 92 °C according to program 9N. The treatment of the textiles was performed by padding and nipping in a Mathis HVF lab scale foulard. Fabrics were folded into three layers and treated at a speed of 2 m/min. The pressure between rollers was set to 3.5 bar for CO, 4 bar for PES, and 2 bar for CO/PES. The bath volume was calculated according to each fabric's water uptake with 10% added. The fabrics were subsequently flat dried at 50 °C in a drying cabinet.

2.2.3. Inter-Fiber Cohesion Measurement

Ten 250 × 100 mm specimens were cut from each web with the fibers positioned lengthwise. The specimen weight was 3.3 g (±0.4 g). Each specimen was conditioned in 65% humidity and 20 °C for 24 h before testing and weighed before testing for normalization of the tensile test result. Tensile tests were performed on a 3 kN Mesdan lab tensile tester with a load cell of 100 N and pneumatic yarn grips. Starting gauge length was 75 mm and rate of extension was 300 mm/min.

The tensile test was performed along the fiber direction, which means that maximum force, F_{max} (N), represent the cohesion between fibers. The maximum force was normalized by the web strip mass, m (g), to obtain cohesion force (CF) in accordance with Equation (1). Nine replications were made.

$$CF = \frac{F_{max}}{m} \quad (1)$$

2.2.4. Textile Tearing

The shredding process took place at RISE IVF, Mölndal, Sweden. The untreated and treated fabrics were cut into smaller pieces in fabric cutter NSX-QD350 and then fed twice into a shredder consisting of four drums. The first drum was NSX-FS1040 with 8 mm long saw teeth, and the subsequent three drums were of type NSX-QT310 with 4 mm saw teeth. The fabric cutter and shredder were from the manufacturer New Shun Xing Environmental Technology.

2.2.5. Fiber Analysis

The recycled fibers were analyzed to measure the efficiency of the tearing process with different pre-treatments of fabrics. Firstly, neps and unopened threads were manually separated from the bulk, and weight percentage was calculated. Secondly, the fiber length was measured by image analysis. This method was chosen to include all lengths of fibers, especially short fibers, which can often be missed in most other fiber measurement methods [26]. A random sample of fibers was carded by hand, and 0.02 g of these were then carefully aligned and placed on green paper. A high resolution picture was taken, and from this picture the fiber length was determined by image analysis (Figure 2). This is a well-established method [26–28].

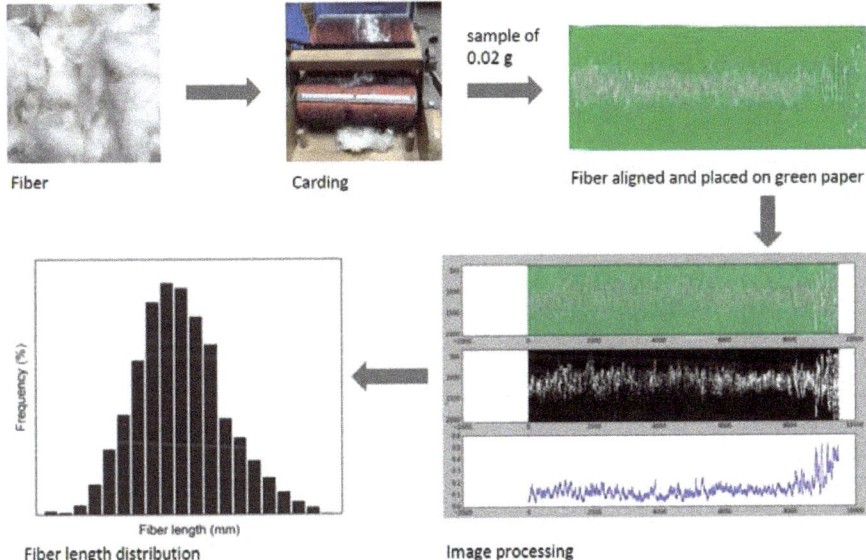

Figure 2. Fiber length measurement method.

The mean length of the fibers, l_m, was determined by the following equation, as found in the literature [29] (p. 138):

$$Mean\ length\ of\ the\ fibers\ (l_m) = \frac{\sum f_i l_i}{\sum f_i} = \frac{\sum f_i l_i}{N} \quad (2)$$

In the equation, l_m is the mean length of the fibers, l_i is fiber length (mm), f_i is number of fibers with the same length, and N is the total number of fibers.

2.2.6. Yarn Spinning and Testing

The recovered fibers from the textile tearing process were carded in the above mentioned carding machine and then drawn into slivers in a Mesdan 3371 stiro-roving lab machine, ready for the rotor machine. The total draft ratio was 1.43.

The rotor spinning of the recycled fibers was performed at RISE IVF in Mölndal on a SDL Atlas Quickspin lab scale spinning machine. A rotor diameter of 40 mm was used together with an OS21 opening roller. The yarn tenacity was tested on a Mesdan lab tensile tester according to standard SS-ISO 2062. Residual lubricant in the yarn is likely to affect the inter-fiber cohesion and thereby also the yarn tenacity. Hence, the yarns were tensile tested, both as received and with the lubricant rinsed away. The yarn linear density was measured.

2.2.7. Statistical Analysis

The CF test and yarn tensile test data were analyzed for statistical significance. The software Minitab 17 was used to perform one-way and two-way ANOVA, and Tukey Pairwise Comparisons and Tukey simultaneous tested the differences of means. A significance level of 0.05 was applied.

3. Results and Discussions

The main drawback of mechanical recycling of textiles is the loss of fiber length. During the recycling process, the fiber interlocking within and between yarns cause frictional forces to break the fibers rather than disentangling it. The aim was to reduce the cohesion with the use of a lubricant pre-treatment and thereby retain fiber length. Further, a method was developed to test fiber cohesion in order to predict the effectiveness of the tearing process.

3.1. Inter-Fiber Cohesion Measurement

Cohesion test result for CO, PES, and CO/PES fiber webs with different treatment concentrations of PEG (Table 1) are shown in Figure 3. The trend is that the CF decreases at 0.25 wt.%, after which the cohesion seems to increase for PES and stay at approximately the same level for CO before increasing at 0.75 wt.%. For PES fibers, the cohesion decreases at 0.75–1.00 wt.% before increasing again. For both PES and CO, the CF increase at 1.25 wt.%. The trend for CO/PES is a decrease in cohesion with the lowest point at 0.75 wt.%. Previous research on how lubricant concentrations affect the friction or cohesion of fibers show that, after an initial increase, friction decreases and is subsequently followed by a further increase [20]. Another work found a minimum friction value at low concentration [21]. The low concentration minimum is explained by a mono-layer of lubricant filling the grooves of the fiber [20,21]. With higher concentration, the lubricant sticks to itself between fibers, increasing the cohesion between fibers [20]. However, to our knowledge, there is no research that describes the effects of PEG concentration dependence on fiber-fiber interactions.

The ANOVA analysis on CF results showed that there are significant differences between one or more treatment concentrations for all fiber types. Tukey pairwise comparisons showed that, for CO, 1.25 wt.% is significantly different from all other concentrations, and for PES, 0.25 and 1.25 wt.% are the only concentrations significantly different from each other. For CO/PES, 0.75 and 1.25 wt.% are significantly different from untreated fiber webs, and 0.5 and 0.75 wt.% are significantly different from each other. The results show that there are trends and some significant differences; however, it will be the analysis of recycled fibers that shows how the lubricant treatment affects the recycling process.

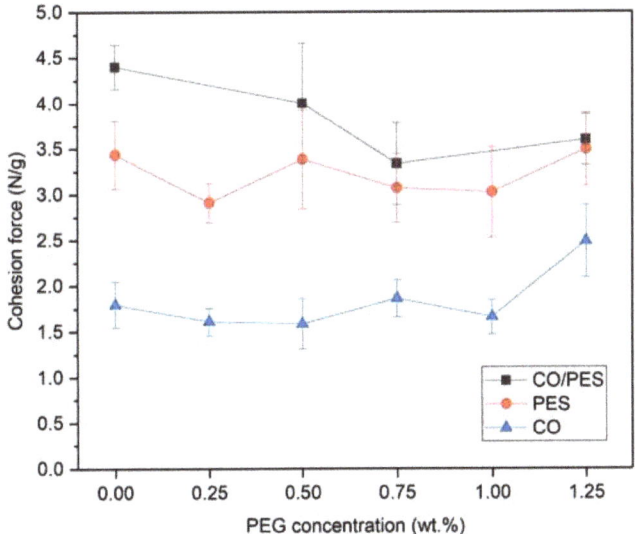

Figure 3. Results from developed inter-fiber cohesion test. CF of CO, PES, and CO/PES fiber webs.

3.2. Recovered Fiber Analysis

Recycling was performed on three samples of each fabric material; one test for non-treated fabric and two with different pre-treatment PEG concentrations. This was to quantify the effect of pre-treatment on the quality of recycled fibers. A low concentration of PEG 4000 (0.1 or 0.2 wt.%) was chosen for the three types of fabrics to identify the effect of a small amount of lubricant. Further, a second higher pre-treatment concentration value was chosen where we tried to achieve the optimal value (minimum CF) obtained from the results of inter fiber cohesion measurements presented in Figure 3. Unfortunately, there were some unforeseen experimental difficulties with the CO/PES material, which is why there is a difference between the higher pre-treatment concentration and the lowest CF detected; see Table 3 for concentrations on pre-treatments on fabrics.

Table 4 shows the result of the analysis on the recycled fibers. A higher concentration of PEG gave a decreased fraction of unopened fiber and preserved fiber length for all fabrics. The decrease of neps and threads is shown to be 21% for CO, 50% for PES, and 18% for CO/PES. As the presence of neps is an indication of fiber breakage and mechanical stresses [30], the recycling process for treated fabrics is shown to be gentler.

The length of recycled cotton fibers was 4.3 mm longer for higher concentration treated CO fiber compared to untreated fabric. For PES, the fiber length difference was even higher; fabric treated with 0.7 wt.% PEG gave 9.4 mm longer fibers compared to untreated fabrics. Recycled CO/PES fabric treated with 0.5 wt.% PEG gave 3.7 mm longer fibers compared to untreated fabrics.

For all fabric materials, the tearing process was gentler at the highest PEG loading. Additionally, for PES, there were areas of melted fibers in untreated recycled fabric, which completely disappeared for the treated fractions. This could be seen as evidence that there is in fact a decrease of cohesion during the tearing process for lubricant treated fabrics.

The average length of these fibers increased on PEG treatment of fabrics before the tearing process, as seen in Table 4. The fiber length distribution for recycled fibers can be seen in Figure 4. The concentration of PEG affected the fiber length. The change was most pronounced in PES, and it can be seen that the treatment with 0.7 wt.% PEG gave fibers as long as 28 mm, while the longest for the 0.2 wt.% PEG was 20 mm and 10 mm for the untreated fabric. Fiber length distribution was also generally wider after the treatment. This could be due to the fact that the fibers in the treated fabric broke less during processing and thus gave a wider range of lengths. These results fall in line with other results in this work.

Table 4. Fiber analysis of the fiber length and neps and unopened threads. Reference values from Table 1.

Material	Pre-Treatment	Neps and Threads Weight %	Mean Fiber Length (mm)
CO	Reference		20.6/27.6
	0.0 wt.% PEG	23.2%	9.1
	0.1 wt.% PEG	21.2%	11.9
	0.3 wt.% PEG	18.4%	13.4
PES	Reference		-/28.0
	0.0 wt.% PEG	44.4%	7.8
	0.2 wt.% PEG	41.2%	15.1
	0.7 wt.% PEG	22.0%	17.2
CO/PES	Reference		25.4/28.8
	0.0 wt.% PEG	22.4%	9.4
	0.1 wt.% PEG	22.4%	12.9
	0.5 wt.% PEG	18.4%	13.1

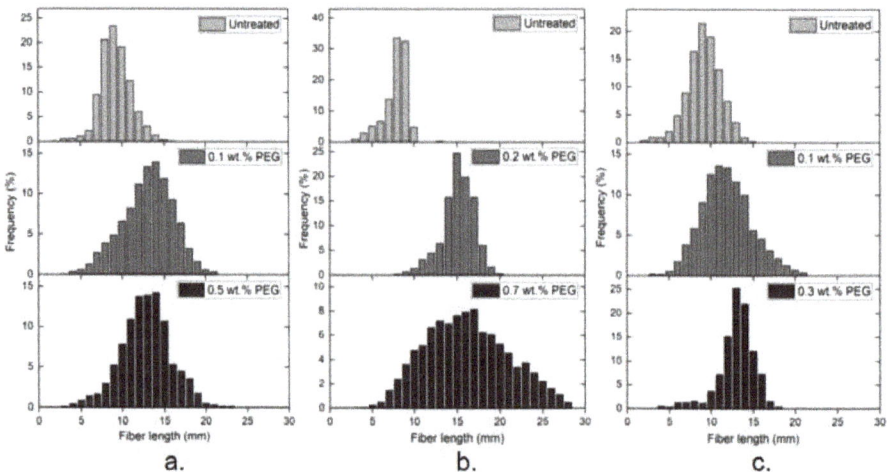

Figure 4. Fiber length distributions for (a) CO, (b) PES, and (c) CO/PES recycled fibers.

Effect of lubricant on the efficiency of the tearing process was evident on CO and PES fiber samples, which is shown in Figure 5. This representation of the recycled fibers shows visually what has been shown in numbers in Table 4 and fiber length distribution in Figure 4. Especially for PES, the length of fibers increased with a higher loading of PEG. In Figure 5a,b, it can be seen that the treated samples are more wooly, less dense and less threads and neps are visible. The CO/PES fiber samples also show a difference between treatments in Figure 5c, albeit a less distinct change.

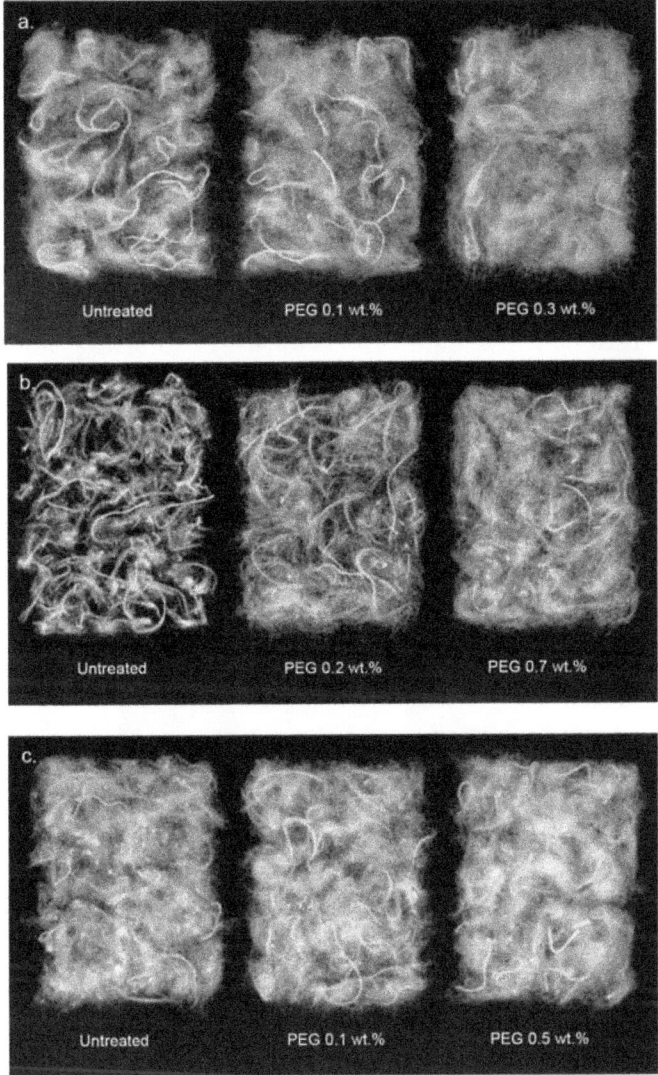

Figure 5. Recycled fibers, 0.05 g of (**a**) CO, (**b**) PES, and (**c**) CO/PES.

3.3. Yarn Spinning

The possibility to rotor spin yarn from 100% recycled fibers was examined to further study the effect of the pre-treatment by a lubricant. During the preparation, it was discovered that it was not possible to process the recycled fibers from CO/PES and untreated PES in the drawing frame, as it was not possible to attain an even sliver. For the CO/PES this can be explained by the high difference in fiber length between the recycled CO and PES. For untreated PES, carding difficulties are related to neps of partly molten fibers that appear to have fused together in the recycled fibers.

During the manufacturing of yarn from recycled CO and PES, it was noted that most neps and threads were removed during the production of sliver and during rotor spinning. This gave the positive outcome that the yarn quality was largely unaffected by the amount of unopened material.

The PES fibers treated with 0.2 wt.% PEG were difficult to spin into yarn, which is why we chose a relatively high linear density for all yarns. The linear densities of the yarns are shown in Table 5.

Table 5. Linear density of rotor spinning yarn.

	CO			PES		
PEG conc. (wt.%)	0.0	0.1	0.3	0.0	0.2	0.7
Yarn linear density (tex)	99	98	92		96	104

Tensile test results for rotor spun yarn of recycled CO are presented in Figure 6. It can be seen that the tenacity of 100% recycled CO rotor spun yarns decrease with the increased concentration of lubricant pre-treatment even though the fiber length was better preserved. For the recycled PES yarns (Figure 7), the tenacity increase with the fiber length, even though the concentration of lubricant is increased.

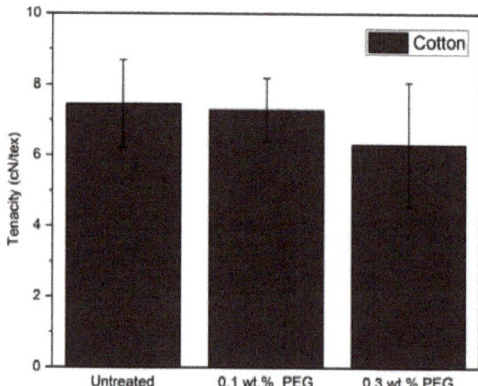

Figure 6. Tenacity of rotor spun yarn from 100% recycled fibers for CO.

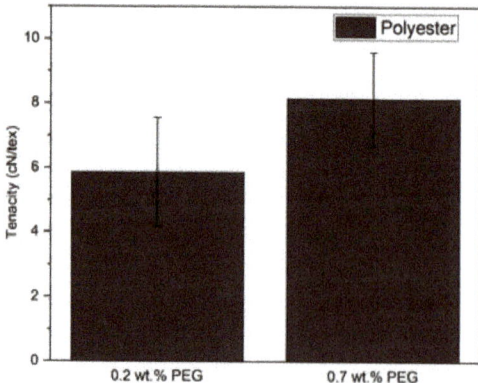

Figure 7. Tenacity of rotor spun yarn from 100% recycled fibers for PES.

To understand the effect of lubricant in the tenacity of recycled yarn, the yarns were washed and the tenacity was retested after washing. Figure 8 show that the tenacity of each yarn is higher after washing. Due to the difficulties in producing yarn from PES treated with 0.2 wt.% PEG, the quantity of yarn was not enough to test the tenacity after washing.

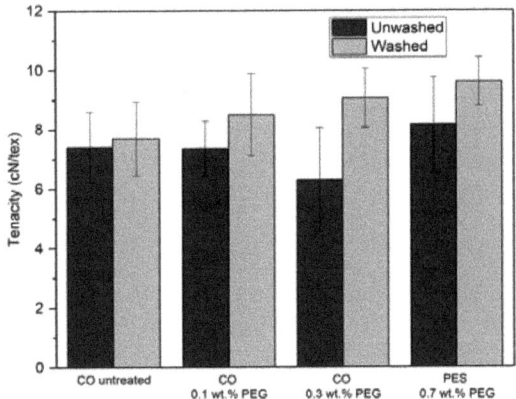

Figure 8. Tenacity of rotor spun yarn from 100% recycled fibers; before and after washing of the yarn.

When analyzing this result statistically, ANOVA showed that the pre-treatment did not have a significant effect on CO yarn tenacity, while the washing did. Further analysis with the Tukey Simultaneous test showed that the tenacity was significantly different after washing for all samples except for untreated CO. Further, neither treated yarns were significantly different from untreated yarn before washing.

After washing, only the 0.3 wt.% for CO had significantly higher tenacity than the untreated samples. For PES yarns, ANOVA showed that both washing and pre-treatment concentration had a significant effect on the tenacity. The Tukey Simultaneous test confirmed this.

The yarns containing the lubricant PEG had lower tenacity, while the same yarns showed higher strength after washing. This shows that the presence of PEG alters the mechanical properties of the yarn as normally longer fibers give a stronger yarn. This is explained by the lubricant effect of PEG; PEG decreases the cohesion between fibers and thus the strength of the yarn. However, during washing, PEG is removed and the fiber length influences the yarn strength positively. After washing, the fiber cohesion between the longer fibers of pre-treated recycled fibers gave strength to the yarns.

4. Conclusions

In this paper, we investigated a method to preserve the fiber length upon mechanically recycling fibers using PEG 4000 treatment. The lubricant PEG reduced cohesion between fibers in the cohesion test and in the tearing process. The fabrics disassembled more easily, and the effect was visible on the recycled fibers. The inter-fiber cohesion test proved successful in predicting a more efficient tearing process with lubricant treated fabrics. Pre-treatment with PEG resulted in:

- Decreased inter-fiber cohesion;
- A tearing process with higher efficiency;
- Decreased fiber length reduction during tearing;
- Enabling rotor spun yarn from 100% recycled fibers.

The lubricating effect on yarn tenacity was also shown by the increase of strength after removal of lubricant through washing. This further proves the inter-fiber cohesion reduction effect of PEG.

This paper shows the potential of lubricant treatment to decrease inter-fiber cohesion during tearing and to increase the value of mechanically recycled fibers. As textile waste can constitute many different fiber types, in the future it would be valuable to study suitable lubricants for different fibers and different fiber blends.

Author Contributions: Conceptualization, K.L., T.S., A.P., and N.K.; data curation, K.L. and T.S.; formal analysis, K.L., T.S., A.P., and N.K.; funding acquisition, A.P. and N.K.; investigation, K.L. and T.S.; methodology, K.L., T.S., A.P., and N.K.; project administration, K.L. and T.S.; resources, A.P.; software, N.K.; supervision, A.P. and N.K.; validation, K.L., A.P., and N.K.; visualization, K.L. and T.S.; writing—original draft, K.L.; writing—review and editing, K.L., T.S., A.P., and N.K. All authors have read and agreed to the published version of the manuscript.

Funding: This work was part of a project funded by Region Västra Götaland.

Acknowledgments: This work was made in collaboration with RISE IVF.

Conflicts of Interest: The authors declare no conflict of interest.

References

1. Sandin, G.; Peters, G. Environmental Impact of Textile Reuse And Recycling—A Review. *J. Clean. Prod.* **2018**, *184*, 353–365. [CrossRef]
2. Ellen MacArthur Foundation, Circular Fashion—A New Textiles Economy: Redesigning Fashion's Future. 2017. Available online: https://www.ellenmacarthurfoundation.org/publications/a-new-textiles-economy-redesigning-fashions-future (accessed on 20 August 2020).
3. Aronsson, J.; Persson, A. Tearing of Post-Consumer Cotton T-Shirts and Jeans of Varying Degree of Wear. *J. Eng. Fibers Fabr.* **2020**, *15*. [CrossRef]
4. Merati, A.A.; Okamura, M. Producing Medium Count Yarns from Recycled Fibers with Friction Spinning. *Text. Res. J.* **2004**, *74*, 646–651. [CrossRef]
5. Wanassi, B.; Azzouz, B.; Ben Hassen, M. Value-Added Waste Cotton Yarn: Optimization of Recycling Process and Spinning of Reclaimed Fibers. *Ind. Crop. Prod.* **2016**, *87*, 27–32. [CrossRef]
6. Taher, H.M.; Bechir, A.; Mohamed, B.H.; Faouzi, S. Influence of Spinning Parameters and Recovered Fibers From Cotton Waste on the Uniformity and Hairiness of Rotor Spun Yarn. *J. Eng. Fibers Fabr.* **2009**, *4*, 36–44.
7. Duru, P.N.; Babaarslan, O. Determining an Optimum Opening Roller Speed for Spinning Polyester/Waste Blend Rotor Yarns. *Text. Res. J.* **2003**, *73*, 907–911. [CrossRef]
8. Khan, K.R.; Hossain, M.M.; Sarker, R.C. Statistical Analyses and Predicting the Properties of Cotton/Waste Blended Open-End Rotor Yarn Using Taguchi OA Design. *Int. J. Text. Res.* **2015**, *4*, 27–35.
9. Nikonova, E.A.; Pakshver, A.B. The Friction Properties of Textile Yarns. *Fibre Chem.* **1973**, *4*, 657–660. [CrossRef]
10. Campos, R.; Bechtold, T.; Rohrer, C. Fiber Friction in Yarn—A Fundamental Property of Fibers. *Text. Res. J.* **2003**, *73*, 721–726. [CrossRef]
11. Waggett, G. The Tensile Properties of Card and Drawframe Slivers. *J. Text. Inst. Trans.* **1952**, *43*, T380–T395. [CrossRef]
12. Ghosh, S.; Rodgers, J.E.; Ortega, A.E. RotorRing Measurement of Fiber Cohesion and Bulk Properties of Staple Fibers. *Text. Res. J.* **1992**, *62*, 608–613. [CrossRef]
13. Deluca, L.B.; Thibodeaux, D.P. The Relative Importance of Fiber Friction and Torsional and Bending Rigidities in Cotton Sliver, Roving, and Yarn. *Text. Res. J.* **1992**, *62*, 192–196. [CrossRef]
14. Barella, A.; Sust, A. Cohesion Phenomena in Cotton Rovings and Yarns: Part I: General Study. *Text. Res. J.* **1962**, *32*, 217–226. [CrossRef]
15. Barella, A.; Sust, A. Cohesion Phenomena in Cotton Rovings and Yarns: Part III: Influence of Fiber Characteristics on the Cohesion of Nontwisted Slivers. *Text. Res. J.* **1964**, *34*, 283–290. [CrossRef]
16. Subramaniam, V.; Sreenivasan, K.; Pillay, P.R. Studies in Fibre Friction: Part I-Effect of Friction on Fibre Properties and Processing Performance of Cotton. *Ind. J. Text. Res.* **1981**, *6*, 9–15.
17. ASTM D2612-99(2018) Standard Test Method for Fiber Cohesion in Sliver and Top (Static Tests). Available online: https://www.astm.org/Standards/D2612.htm (accessed on 2 July 2020).
18. Scardino, F.L.; Lyons, W.J. Influence of Fiber Geometry on the Mechanical Properties of Assemblies during Processing: Part I: Polyester Fibers in Webs and Slivers. *Text. Res. J.* **1970**, *40*, 559–570. [CrossRef]
19. Olsen, J.S. Measurement of Sliver Drafting Forces. *Text. Res. J.* **1974**, *44*, 852–855. [CrossRef]
20. Berberi, P.G. Effect of Lubrication on Spinning Properties of Dyed Cotton Fibers. *Text. Res. J.* **1991**, *61*, 285–288. [CrossRef]
21. Olsen, J.S. Frictional Behavior of Textile Yarns. *Text. Res. J.* **1969**, *39*, 31–37. [CrossRef]

22. Ajayi, A. Friction in Woven Fabrics. In *Friction in Textile Materials*; Gupta, B.S., Ed.; Woodhead Publishing Limited: Cambridge, UK, 2008; pp. 351–385.
23. Stepanova, T.Y. The Effect of Lubricants on Tribological Characteristics of Fibrous Materials. *J. Frict. Wear* **2016**, *37*, 430–434. [CrossRef]
24. Kobayashi, M.; Koide, T.; Hyon, S.H. Tribological Characteristics of Polyethylene Glycol (PEG) as a Lubricant for Wear Resistance of Ultra-High-Molecular-Weight Polyethylene (UHMWPE) in Artificial Knee Join. *J. Mech. Behav. Biomed. Mater.* **2014**, *38*, 33–38. [CrossRef] [PubMed]
25. Zarrintaj, P.; Saeb, M.R.; Jafari, S.H.; Mozafari, M. Application of Compatibilized Polymer Blends in Biomedical Fields. *Compat. Polym. Blends* **2019**, 511–537. [CrossRef]
26. Ikiz, Y.; Rust, J.P.; Jasper, W.J.; Trussell, H.J. Fiber Length Measurement by Image Processing. *Text. Res. J.* **2001**, *71*, 905–910. [CrossRef]
27. Pinter, P.; Bertram, B.; Weidenmann, K.A. A Novel Method for the Determination of Fibre Length Distributions from μCT-data. In Proceedings of the 6th Conference on Industrial Computed Tomography (iCT) 2016, Wels, Austria, 9–12 February 2016.
28. Wang, H. *Fiber Property Characterization by Image Processing*; Texas Tech University: Lubbock, TX, USA, 2007.
29. Hearle, J.; Morton, W. *Physical Properties of Textile Fibres*, 4th ed.; Woodhead Publishing Limited: Cambridge, UK, 2008.
30. Van Der Sluijs, M.H.J.; Hunter, L. A Review on the Formation, Causes, Measurement, Implications and Reduction of Neps during Cotton Processing a Review on The Formation, Causes, Measurement, Implications and Reduction of Neps During Cotton Processing. *Text. Prog.* **2016**, *48*, 221–323. [CrossRef]

Publisher's Note: MDPI stays neutral with regard to jurisdictional claims in published maps and institutional affiliations.

© 2020 by the authors. Licensee MDPI, Basel, Switzerland. This article is an open access article distributed under the terms and conditions of the Creative Commons Attribution (CC BY) license (http://creativecommons.org/licenses/by/4.0/).

 *sustainability*

Article

Environmental Profile Study of Ozone Decolorization of Reactive Dyed Cotton Textiles by Utilizing Life Cycle Assessment

Ajinkya Powar [1,2,3,4,*], Anne Perwuelz [1,*], Nemeshwaree Behary [1], Le Vinh Hoang [5], Thierry Aussenac [5], Carmen Loghin [3], Stelian Sergiu Maier [3], Jinping Guan [4] and Guoqiang Chen [4]

1. Ecole Nationale Supérieure des Arts et Industries Textiles (ENSAIT), GEMTEX Laboratory, 59056 Roubaix, France; nmassika.behary@ensait.fr
2. Université de Lille, Nord de France, F-59000 Lille, France
3. Faculty of Industrial Design and Business Management, Gheorghe Asachi Technical University of Iasi, 67, 700050 Iasi, Romania; cloghin@tex.tuiasi.ro (C.L.); smaier@ch.tuiasi.ro (S.S.M.)
4. College of Textile and Clothing Engineering, Soochow University, Suzhou 215000, China; guanjinping@suda.edu.cn (J.G.); chenguojiang@suda.edu.cn (G.C.)
5. Institut Polytechnique UniLaSalle, Université d'Artois, ULR 7519, 60026 Beauvais, France; Levinh.Hoang@unilasalle.fr (L.V.H.); thierry.aussenac@unilasalle.fr (T.A.)
* Correspondence: ajinkya.powar@ensait.fr (A.P.); anne.perwuelz@ensait.fr (A.P.)

Citation: Powar, A.; Perwuelz, A.; Behary, N.; Hoang, L.V.; Aussenac, T.; Loghin, C.; Maier, S.S.; Guan, J.; Chen, G. Environmental Profile Study of Ozone Decolorization of Reactive Dyed Cotton Textiles by Utilizing Life Cycle Assessment. *Sustainability* 2021, 13, 1225. https://doi.org/10.3390/su13031225

Academic Editor: Hanna de la Motte
Received: 11 December 2020
Accepted: 20 January 2021
Published: 25 January 2021

Publisher's Note: MDPI stays neutral with regard to jurisdictional claims in published maps and institutional affiliations.

Copyright: © 2021 by the authors. Licensee MDPI, Basel, Switzerland. This article is an open access article distributed under the terms and conditions of the Creative Commons Attribution (CC BY) license (https://creativecommons.org/licenses/by/4.0/).

Abstract: Research approaches on the use of ecotechnologies like ozone assisted processes for the decolorization of textiles are being explored as against the conventional alkaline reductive process for the color stripping of the cotton textiles. The evaluation of these ecotechnologies must be performed to assess the environmental impacts. Partial "gate to gate" Life Cycle Assessment (LCA) was implemented to study the ozone based decolorization process of the reactive dyed cotton textiles. Experiments were performed to determine input and output data flows for decolorization treatment of reactive dyed cotton textile using the ozonation process. The functional unit was defined as "treatment of 40 g of reactive dyed cotton fabric to achieve more than 94% color stripping". Generic and specific data bases were also used to determine flows, and International Life Cycle Data system (ILCD) method was selected to convert all flows into environmental impacts. The impact category "Water resource depletion" is the highest for all the ozonation processes as it has the greatest relative value after normalization amongst all the impact indicators. Electricity and Oxygen formation were found to be the major contributors to the environmental impacts. New experimental conditions have been studied to optimize the impacts.

Keywords: life cycle assessment; normalization method; environmental impacts; ozonation process; decolorization; reactive dyed cotton textiles; "gate-to-gate" life cycle assessment (LCA)

1. Introduction

Reactive dyes constitute nearly 50% of the worldwide market for the coloring of cellulose-based fibers [1]. However, the coloration industry faces some general problems such as faulty or uneven dyeing and the presence of color patches on the surface of the textile fabrics during coloration and subsequent processing operations [2,3].

To rectify these problems, the normal approach practiced in the coloration industry is destructive stripping. But, this traditional and commonly employed technique consists of huge amounts of various oxidizing and/or reducing agents in a single color stripping process, such as hydrogen peroxide, sodium hypochlorite, chlorine dioxide, and dichromate salts, as well as thiourea dioxide, sodium hydrosulphite, and decroline [4]. In addition to this, these traditional color stripping methods involve high temperature and use of chemicals, both of which contribute to ecological pollution loads, huge liquor consumptions, and high production costs [5,6]. Therefore, a sustainable process needs to be implemented to overcome these drawbacks. Previous studies demonstrated the use of the biological stripping method as an ecofriendly and cost effective approach [4,7]. Recent studies showed that the photocatalytic system with UV/$Na_2S_2O_4$, used for color stripping, was more energy effective, ecological, and a sustainable alternative [8]. In our study, we have proposed the use of the ozonation process for the color stripping of the reactive dyed fabric in a pilot scale process. No previous studies are available on the environmental impact assessment of the designed color stripping process.

A significant contribution to environmental pollution and resource utilization has been caused by the textile industry [9]. Therefore, the textile industry (TI) is exploring emerging and cleaner technologies in order to minimize the use of natural resources. Further, the TI seeks to continuously improving sustainable activity techniques, thus aiming for zero emissions [10]. In that context, it is important to reduce the amount of textile waste due to manufacturing defects without increasing the overall impact.

The life cycle assessment (LCA) is defined as a compilation and assessment of the inlets, outlets, and potential environmental impacts of a process or product through its life cycle. It is a vital tool to carry out the environmental analysis [11]. LCA is a model to determine the manufacturing methods either they are sustainable or cyclic production and find a substitute ecofriendly production process. LCA studies are principally determined as "gate-to-gate" or "cradle to grave" [12].

For the color stripping process, the environmental impacts of the proposed ozonation method has to be quantified in order to justify the profile of the process [13]. In our study, we used gate-to-gate LCA methodology. The detailed color stripping and mechanical property characterization has already been discussed in a previous paper [14]. The environmental assessment in our work is based on defining a functional unit color stripping of 40 g of reactive dyed cotton fabric to achieve color removal, and the determination of different environmental impact categories for the ozone assisted color stripping method.

The aim of the study was to highlight the main contributors to the environmental impact of the ozone stripping process and then to find the best conditions for reactive dyed decoloration. In addition, this work intends to identify and evaluate the potential impact of the ozonation process used and also encourages the sustainability profile of the process.

2. Experimental Work

2.1. Woven Cotton Textile

A 100% Cellulosic (cotton) woven fabric (150 g/m^2) was implemented in this study. The cotton fabric was dyed with a 1% reactive dye (C.I. Reactive Black 5). This dyed cotton fabric was used for the decolorization treatment of the fabric.

2.2. System Considered: Color Stripping Using the Ozone-Assisted Treatment

The ozone-assisted process was carried out using a pilot scale ozone reactor at the Unilasalle laboratory, France. The ozonation system is described in Figure 1 [14].

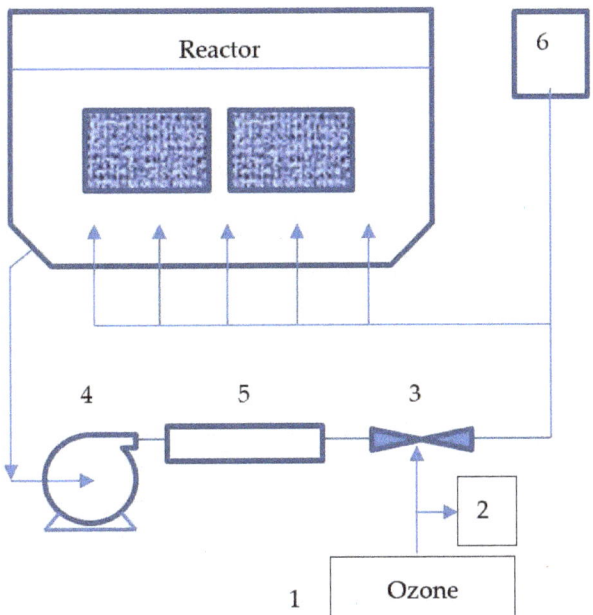

Figure 1. Pilot ozonation (1, ozone generator; 2, analyser ozone; 3, venturi injection system; 4, circulation pump; 5, filter; 6, dissolved ozone analyzer and pH meter) [14].

Ozone is produced by the electric discharge in oxygen provided by liquid oxygen pressurized bottles. Ozone transfer from the gas phase to the liquid phase is an important process to obtain the dissolved ozone in water in the reactor. Various techniques of the gas dispersion are applied in practice and diffusers, static mixer, injection etc. are the most popular ones [15]. In this study, we used the venturi injection process.

The oxygen O_2 flow rate is constant at $F = 0.3$ m^3/h, and the amount of ozone is measured in situ. The excess ozone is then destroyed in a 0.8 kW ozone destructor ODT-003.

The water bath used was at a fixed volume of 60 L of tap water. The circulation pump of the reactor has a power of 0.75 kW, and we made the assumption that only 10% of the power is required. All the experiments are made at room temperature. The pH value was regulated by adding phosphoric acid (PanReac AppliChem) and sodium hydroxide (EMPLURA® Merck, Germany). The pH was measured in situ during the ozonation process.

A Box Behnken experimental design matrix was setup to find the best experimental conditions for the decolorization with varying ozone concentration, pH, and treatment time, as described in our previous paper [14].

The 40 g blue dyed cotton fabric was placed in the reactor and subjected to the ozone treatment. As a result, the treated fabric started to decolorize and the color stripping % was measured using a spectrocolorimeter (Figure 2).

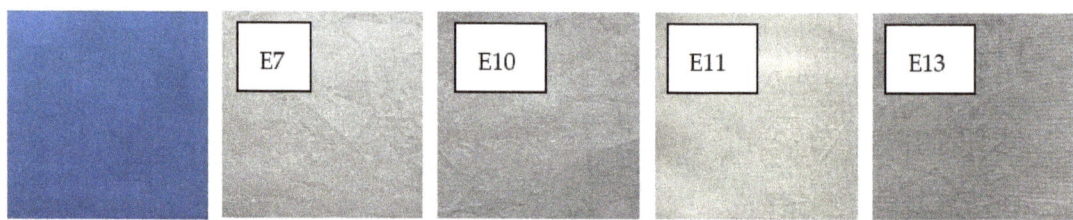

Figure 2. Cotton sample before and after ozonation treatment [14].

In this paper, we only considered experiments with stripping values more than 94% (Table 1). The best stripping was obtained with experiment E11 performed at pH 5. It had an ozone concentration in the oxygen gas flow of 85 g/m³ NTP (normal pressure and temperature) and a treatment time of 50 min. We considered the E11 experiment as a reference. In the experiments E7 and E8, the treatment time decreased to 30 min only, while a lower ozone concentration was used in E10 and E12. The E13–E16 experiments all had less ozone and less time, yet the stripping results were not good as compared to the reference.

Table 1. Ozonation experimental conditions with color stripping %.

Sr. No.	pH	Concentration Ozone (g/m³ TPN)	Time (min)	Color Stripping %
E7	7	85	30	95.1
E8	3	85	30	97.45
E10	7	45	50	94.3
E11	5	85	50	97.6
E12	3	45	50	97.5
E13	5	45	30	94.6
E14	5	45	30	94.1
E15	5	45	30	94.65
E16	5	45	30	93.9

2.3. Material and Energy Requirement

The amount of resources required for the treatment of 40 g reactive dyed fabric was estimated from the treatment parameters and the characteristics of the devices of the process (Table 2).

Table 2. Ozone and energy requirements reference process E11.

Sr. No.	Inputs from Technosphere	Quantity
1	Energy for ozone generation with plasma treatment (kWh)	0.213
2	Energy for the circulation pump (kWh)	0.0625
3	Energy for the ozone destructor ODT-003 (kWh)	0.077
4	Oxygen (Kg)	0.357

a. Oxygen O_2 and ozone O_3 requirements:

The amount of O_2 required was calculated from the flowrate and the treatment time. For the reference process E11, the treatment time was 50 minutes. So the amount of O_2 required was 0.25 m³ corresponding to 0.357 kg of Oxygen as the oxygen 'O_2' density is 1.429 kg/m³.

The concentration of O_3 in the oxygen flow was constant. Thus, the total amount of O_3 produced was calculated from the volume of oxygen 'O_2' gas used. With the oxygen concentration of 0.85 g/m³, the O_3 amount produced was equal to 21.25 g.

b. Energy requirements: Energy-associated concerns:

- Ozone generation with plasma treatment: Specific energy required to produce one kg of ozone from liquid oxygen was 7–13 kWh/kg O_3 [16]. An average value of 10 Wh/g O_3 was selected for our study, and thus this energy in the reference experiment was 212.5 Wh.
- Ozone destructor ODT-003 operated at a power of 0.8 kW, which was associated with the maximum gas flow rate of 3.7 kg/h [16], or 2.59 m^3/h with oxygen gas. As we used only 0.3 m^3/h, then the power needed is 0.092 kW which when multiplied by the treatment time, yields the quantity of energy used. With experiment E11, which was carried for 50 min, the ozone destructor energy was 77 Wh.
- Water circulation pump of the reactor: Multiplying the 0.075 kW power by the treatment time provided the energy used, and for the E11 experiment, it was 62.5 Wh.
- For the reference treatment, E11, the total electricity requirement was 0.352 kWh.

c. Chemicals

The water bath was made with tap water. In case of the reference process at pH = 5, the amount of phosphoric acid and sodium hydroxide used were 6.75 and 3.65 g, respectively.

3. Life Cycle Assessment

The LCA was modeled with the SIMAPRO LCA software tool as per the international standard. The decolorization or color stripping of textiles is a unit process carried out in the textile production value chain to rectify the faults or unevenness issues occurred during cotton textile manufacturing. "Gate-to-gate" LCA analysis considers only the color stripping process to study the environmental profile of the ozone-assisted process. The method used for the assessment of the environmental impacts was from International Reference Life Cycle Data System ILCD 2011 Midpoint+ V1.07/EU27 2010, equal weighting.

From the 16 impact categories of the ILCD method, the 6 following have been reported: climate change; water resource depletion; human toxicity (cancer effects); freshwater ecotoxicity; mineral, fossil, and ren resource depletion; and the ionizing radiation of human health (HH).

To compare the significance of each impact category, they were all normalized using the 2010 normalization factors related to the EU-27 impacts [17]. In this study, the environmental impacts of a european person annually in 2010 are concerned.

3.1. Goal and Scope Definition

The functional unit was defined as treating "40 g of dyed cotton fabric to achieve specific decolorization". Various process scenarios leading to dyed fabric decolorization were studied for ozone-based color stripping processes. For the proposed ozone-based color stripping process, the "gate-to-gate" system boundaries considered the decolorization step for the manufacturing of chemicals and electrical energy (see Figure 3). Dyed woven cotton fabric manufacturing and chemical transportation were excluded. We hypothesized that the color stripping process was carried out in France. Production sites of chemicals and energy were in Europe. The following elements were outside the system boundaries: transport of chemicals and the fabrication/maintenance of the ozone machine and wastewater treatment. In this model, tap water was considered to minimize the impacts due to the use of deionized water or reverse osmosis water. There may be slight variations in the actual results due to the use of tap water. Moreover, the catalytic process was utilized for the ozone destruction, and hence the output was considered in terms of energy utilized for the destruction of the leftover ozone. Drying the samples was excluded from this study as it did not differ from one treatment to another.

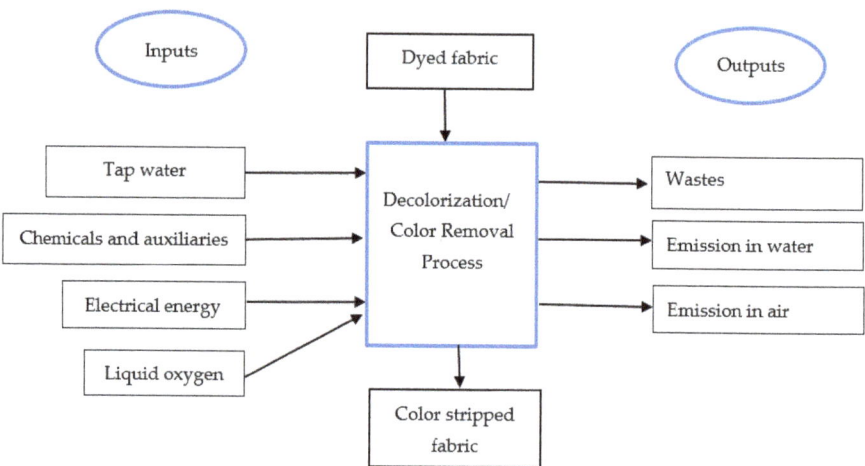

Figure 3. System boundaries of the decolorization process of the reactive dyed cotton.

3.2. Life Cycle Inventory

For the ozone-assisted process, the experimental data was used considering the pilot scale designed machine. The scenarios were determined via laboratory experiments. From these scenarios, data were obtained to quantify flow inputs (consumed resources) and outputs (emissions or outcomes of the process). In our studies, data were obtained from several sources (Table 3). Specific data from experiments carried out in the laboratory and the production data was collected from the ECO INVENT database. These inventory data included the production of chemicals, liquid oxygen, and tap water in Europe (RER datasets), as well as electricity production and distribution in France (FR datasets).

Table 3. The life cycle inventory for the decolorization of 40 g of reactive dyed cotton fabric using the ozonation technique (reference the E11 experiment).

Inputs	Unit	Amount	Description	Source
Phosphoric acid	g	6.76	Phosphoric acid, industrial grade, without water, in 85% solution state {RER} | purification of wet-process phosphoric acid to industrial grade, product in 85% solution state | Alloc Rec, S	Eco-invent database
Tap water	mL	60,000	Tap water {Europe without Switzerland} | tap water production | underground water without treatment | Alloc Rec, S	Eco-invent database
Sodium hydroxide	g	3.55	Sodium hydroxide, without water, in 50% solution state {RER} | chlor-alkali electrolysis, diaphragm cell | Alloc Rec, S	Eco-invent database
Electricity	kWh	0.352	Electricity grid mix, AC, consumption mix, at consumer, 230 V FR, S	Eco-invent database
Oxygen	g	0.357	Oxygen, liquid {RER} | air separation, cryogenic | Alloc Rec, S	Eco-invent database
Outputs	Unit	Amount	Description	
Phosphoric acid	g	6.76	Wastewater content	
Tap water	mL	60,000	Wastewater content	
Sodium hydroxide	g	3.55	Wastewater content	

4. LCA Results

4.1. LCIA Results and Interpretation for the Reference Scenario

The main environmental impacts are described in Table 4. The total greenhouse gas (GHG) produced by the ozone treatment was 213 g of equivalent CO_2. The water depletion was 168 L, while the resource depletion was 14 mg equivalent to Sb. The ionizing radiations were equivalent to 190 becquerel of the U235. The fresh water ecotoxicity was equivalent to 2 comparative toxic units (CTU), while the cancer human toxicity was calculated at 0.02×10^{-6} CTU. The normalization method was added to describe the extent to which the impact categories had a significant influence on the environment [18]. The normalized factor is the environmental impact caused annually by the activities of an average European, it is expressed as "person year equivalent", PEeq.

Table 4. Impact categories and normalized values for the impacts in the ozonation process E11.

Impact Category	Unit	Value (Unit: See Column)	Normalized Value (Unit: PEeq.)
Climate change	kg CO_2 eq	0.21388189	0.0000235
Mineral, fossil, and ren resource depletion	kg Sb eq	0.00001360	0.000135
Ionizing radiation HH	Kbq U_{235} eq	0.19096354	0.000169
Freshwater ecotoxicity	CTUe	2.01311521	0.000229
Human toxicity, cancer effects	CTUh	0.00000002	0.000549
Water resource depletion	m^3 water eq	0.16857161	0.002073

The LCA normalized results for every impact category in the ozone reference process (E11) are displayed graphically in Figure 4, with the same equivalent person year unit. The four major impacts are as follows: water resource depletion, human toxicity, cancer effects, freshwater ecotoxicity, and the ionizing radiation HH. The main environmental impact for the reference process E11 concerned the water resource depletion, as it had the greatest relative value after normalization amongst all of the impact indicators. From Figure 4, we observed that there was a minor impact on climate change, as well as the mineral, fossil, and renewable energy depletion.

Interpretation

Considering the reference E11 ozonation process, we studied the contribution of different materials and electricity for various environmental impacts (Figure 5). We observed that tap water and sodium hydroxide had a negligible share in the environmental impacts. Electricity contributed greatly to the environmental impacts, such as ionizing radiations, water resource depletion, and material depletion. Liquid oxygen contributed greatly to climate change and freshwater ecotoxicity, and, to a lesser extent, ionizing radiation. Phosphoric acid contributed to the human toxicity and freshwater ecotoxicity.

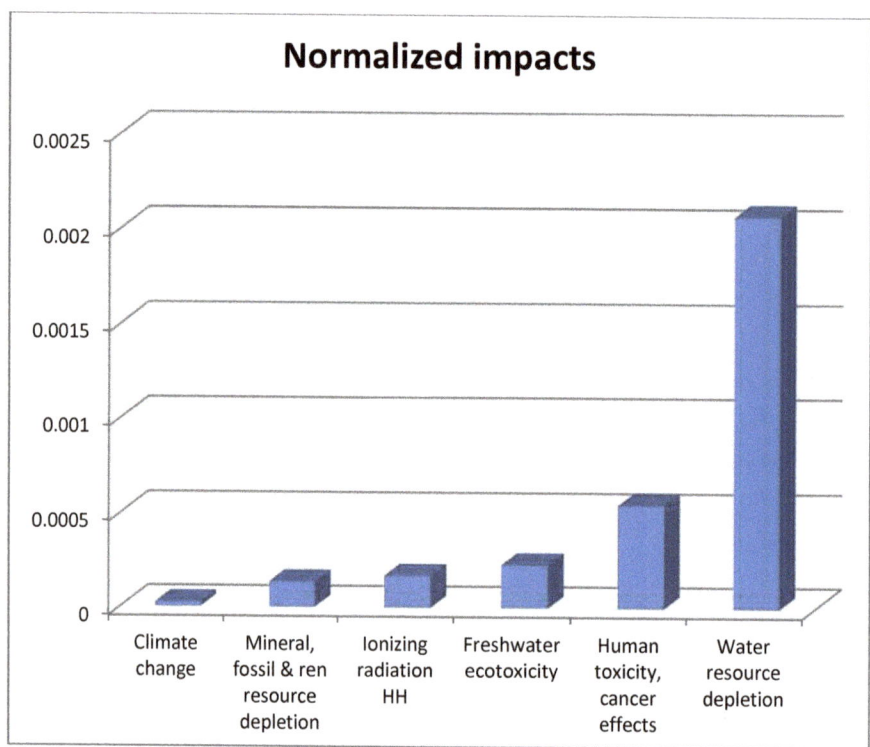

Figure 4. The life cycle assessment (LCA) impact indicators normalized for the ozone-assisted decolorization process E11.

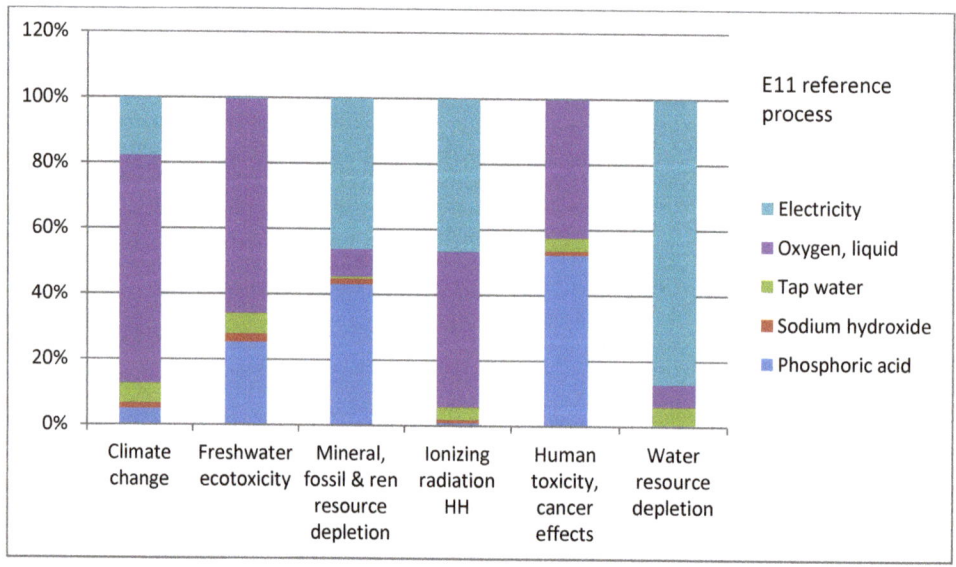

Figure 5. Contribution of different materials and electricity to various environmental impacts: reference process.

Electricity and oxygen formation are the main contributors to environmental impacts. This is related to the ozone generation. Indeed, the main electricity consumption was the ozone generator.

4.2. Process Optimization Regarding Environmental Impacts

As we observed, the environmental impacts were caused by the reference process, and our aim herein was to find the best conditions in terms of the process optimization so that we could minimize such environmental impacts. The inventories for each experiment were calculated according to Section 2.3 (Table 5).

Table 5. Inventories for the ozone experiment.

Sr. No.	O_2 Required	Electricity	Phosphoric Acid	Sodium Hydroxide
	kg	kWh	g	g
E7	0.214	0.211	3.72	1.96
E8	0.214	0.211	6.86	3.6
E10	0.357	0.252	3.72	1.96
E11	0.357	0.352	6.76	3.55
E12	0.357	0.252	6.86	3.6
E13	0.214	0.151	6.76	3.55
E14	0.214	0.151	6.76	3.55
E15	0.214	0.151	6.76	3.55
E16	0.214	0.151	6.76	3.55

When treatment time decreased (Tables 1 and 5), as was the case for experiments E7 and E8, we observed that there was a reduction in the required electricity and O_2 input with very good color stripping.

When the ozone concentration was reduced, such as in experiments E10 and E12 (Tables 1 and 5), we observed that there was reduction in the electricity compared with the reference process. Moreover, we observed very good color stripping.

To take into account both the O_3 concentration decrease and the time reduction, the midpoint experiments of the statistical model (e.g., experiments E13–E16) were selected (Table 5). We clearly observed that the required electricity and O_2 input were less than the reference process. Color stripping was a little bit worst but decolorization still seemed significant.

4.3. Introduction of the LCA Results

Based on the characterized results, we observed that the E13 ozonation process was preferable (Table 6). The largest differences in the impacts were observed between the reference process (E11) and the midpoint of the experiments (E13).

Table 6. Characterization values of the impact categories for the E11 reference, and the E12, E8, and E13 processes.

Impact Category	Unit	E11	E8	E12	E13
Climate change	kg CO_2 eq	0.21388189	0.13921773	0.20321215	0.13248698
Mineral, fossil, and ren resource depletion	kg Sb eq	0.00001360	0.00001072	0.00001191	0.00000956
Ionizing radiation HH	Kbq U_{235} eq	0.19096354	0.11885205	0.16565425	0.10358892
Freshwater ecotoxicity	CTUe	2.01311521	1.48916748	2.01922845	1.47962330
Human toxicity, cancer effects	CTUh	0.00000002	0.00000002	0.00000002	0.00000002
Water resource depletion	m^3 water eq	0.16857161	0.10504289	0.12684691	0.08000251

Figure 6 represents the LCA results for the E11 reference, as well as E12, E8, and E13 processes. Here, the reference ozone process (E11) was compared to the different optimized processes (E8, E12, and E13). Table 6 shows that the optimized processes E8 and E13 had much lower impact values than E12 process for environmental impacts such as climate change, ionizing radiation HH, and water depletion. The reference process E11 had the highest environmental impacts. In our LCA study based on normalized results, the atmospheric impacts, especially water resource depletion, exhibited the poorest performance among every environmental impact category. The reason could be attributed to the ozonation process setup by utilizing a large amount of water. When we observed the midpoint of the experiments, we saw that E13 had lower impacts than the reference, which used less liquid oxygen for the ozone generation and less electricity, thus reducing the overall environmental impacts. However, we obtained less color stripping, as already discussed. (Tables 1 and 6).

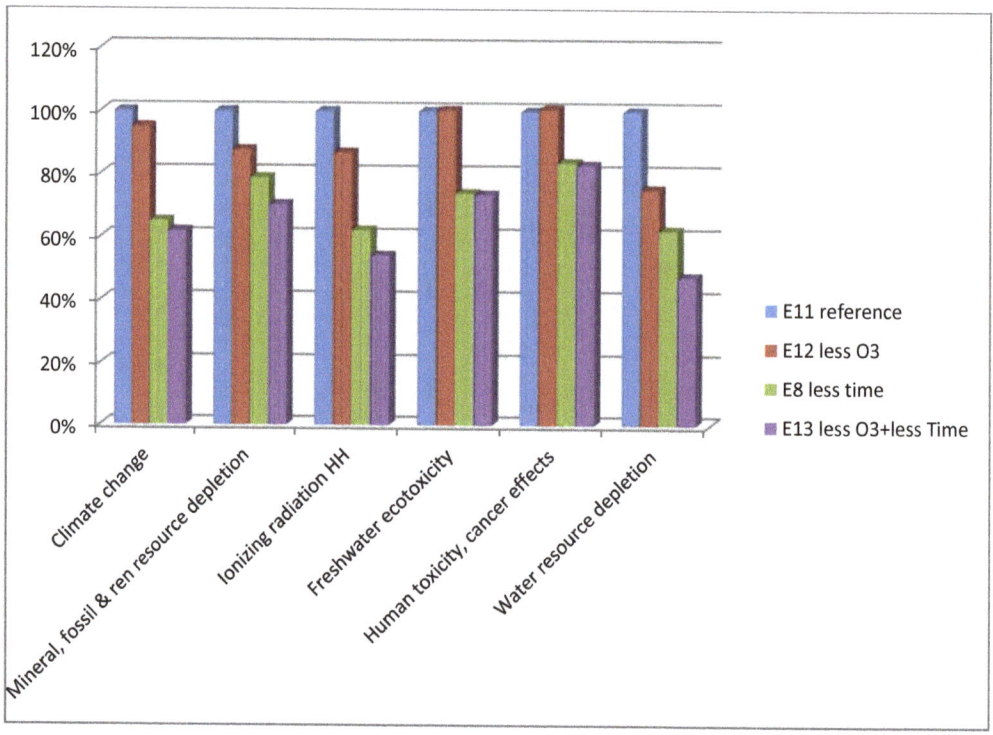

Figure 6. Comparative LCA results (normalized values) for the E11 reference, as well as the E12, E8 and E13 processes.

5. Discussion

So depending on the color specifications, the optimum value could be selected focusing either on the color stripping quality or on the environmental impact. If a color stripping of 94% is enough for example before dark dyeing, then the best conditions would have the lowest impact. The results obtained with the optimum conditions were good and comparable to the literature. Previous studies have shown that the reactive black 5 dyed cotton fabrics were color stripped with 96.1% and 94.4% of the stripping percent, via the electrochemical method [19].

The environmental impact of the ozone-based decolorization process was primarily caused by water use and energy consumption. The reactor utilized operates at higher material to liquor ratios; as the reactor we have used is not dedicated to textiles. The reactor design needs to be improved in order to increase the amount of fabric that could be introduced for the treatment. Since the volume of the water in the reactor was large, this also resulted in the high consumption of chemicals and auxiliaries.

This study shows that electricity is very important. In fact the overall environmental impact depends on the electricity mix and in France the electricity mix has lot of nuclear energy and that's the reason we have high ionizing radiation HH impacts. Impact categories are sensitive to the energy mix of the country. If we change the country with less electricity mix and high carbon content so we have high climate change and less ionizing radiation impact.

Moreover, the reactor utilized large amount of ozone and thus the liquid oxygen which is needed for the production of ozone. Thus, the ozone generator was also a contributor, as discussed previously. In previous studies available on wastewater treatments, the research findings showed that the ozonation process adds a 6% greater impact on climate change. This is attributed to the liquid oxygen and electricity production associated with the ozonation process [20]. In another study on the application of the LCA to the Kraft pulp industrial wastewater treatment via different advanced oxidation processes, it clearly depicted that ozonation accounted for a higher environmental impact, owing to energy consumption produced by the oxygen and ozone [21]. These results coordinate with our study. The results in this study showed that combining the ozonation with UV-A light decreased the environmental impact by about 40% [21]. In a similar study on the analysis of the advanced oxidation process, results showed that high energy consumption was a great drawback in the ozonation process [22].

6. Conclusions

Considering the technique utilized for the decolorization of cotton textiles using the "gate-to-gate" LCA tool, we discerned the environmental profile of the process and the hotspots associated with it.

For the ozone-assisted treatment, the electricity and oxygen formation for ozone generation were major contributors for the environmental impacts. This could be attributed to the ozone generation process, which utilized liquid oxygen and included electricity consumption due to the ozone generator.

The environmental impacts can be reduced with regards to reference process by decreasing the ozone input, decreasing the treatment time and by simultaneously decreasing the treatment time and the ozone input. However, this change in the ozonation parameters had an impact on the color stripping %.

For the ozone-assisted process, energy consumption and wastewater (pollution) related impacts were higher. "Water depletion" and "human toxicity cancer effects" were higher for the selected impact categories. We can reduce the impacts by reducing the liquor use in the ozonation process. The results obtained from the LCA study of the "gate-to-gate" provide necessary solutions that could reduce impacts, find possible solutions, and remodify the technique or process.

This study paves a route to use the ozone-based process for textile processing and allied industries at an industrial scale. It also encourages us to think and develop technologies for industrialists looking for sustainable and environmentally friendly alternatives with lower ecological impacts. Studies on the financial aspects of the process could also be an interesting research area. For future study, the LCA with different textile decolorization methods might be assessed.

Author Contributions: Conceptualization, A.P. (Ajinkya Powar), A.P. (Anne Perwuelz), L.V.H., and T.A.; methodology, A.P. (Ajinkya Powar), A.P. (Anne Perwuelz), and L.V.H.; software, A.P. (Ajinkya Powar), A.P. (Anne Perwuelz); formal analysis, A.P. (Ajinkya Powar) and A.P. (Anne Perwuelz).; data curation, A.P. (Ajinkya Powar), A.P. (Anne Perwuelz), and L.V.H.; writing—original draft preparation, A.P. (Ajinkya Powar); writing—review and editing, A.P. (Ajinkya Powar), A.P. (Anne Perwuelz), N.B., and L.V.H.; supervision, A.P. (Ajinkya Powar), A.P. (Anne Perwuelz), N.B., C.L., S.S.M., J.G., G.C., L.V.H., and T.A. All authors have read and agreed to the published version of the manuscript.

Funding: This research work was realized in the framework of the Erasmus Mundus Joint Doctorate program Sustainable Management and Design for Textiles (SMDTex), which is financed by the European Commission.

Acknowledgments: The author would like to acknowledge the UniLaSalle, Beauvais for their help and support in the ozone experiments and thank Christian Catel from GEMTex Lab at ENSAIT (France). Also the author would like to thank and acknowledge the Teinturerie Lenfant, France, the dye house and the Achitex Minerva, France for their help in the dyeing fabric and also providing the reactive dyestuff respectively.

Conflicts of Interest: The authors declare no conflict of interest.

References

1. Roessler, A.; Jin, X. State of the Art Technologies and New Electrochemical Methods for the Reduction of Vat Dyes. *Dye. Pigment.* **2003**, *59*, 223–235. [CrossRef]
2. Fono, A.; Montclair, N.J. New Process of Color Stripping Dyed Textile Fabric. U.S. Patent No 4,227,881, 14 October 1980
3. Ogulata, R.T.; Balci, O. Investigation of the Stripping Process of the Reactive Dyes Using Organic Sulphur Reducing Agents in Alkali Condition. *Fibers Polym.* **2007**, *8*, 25–36. [CrossRef]
4. Ali, S.; Chatha, S.; Asgher, M.; Ali, S.; Ijaz, A. Biological Color Stripping: A Novel Technology for Removal of Dye from Cellulose Fibers. *Carbohydr. Polym.* **2012**, *87*, 1476–1481. [CrossRef]
5. Long, J.; Liu, B.; Wang, G.; Shi, W. Photocatalitic Stripping of Fi Xed Reactive Red X-3B Dye from Cotton with Nano-TiO_2/UV System. *J. Clean. Prod.* **2017**, *165*, 788–800. [CrossRef]
6. Nalankilli, G. Stripping of Dyes from Faulty Dyeings. *Colourage* **1997**, *44*, 33–39.
7. Chatha, S.A.; Ali, S.; Asgher, M.; Bhatti, H.N. Investigation of the Potential of Microbial Stripping of Dyed Cotton Fabric Using White Rot Fungi. *Text. Res. J.* **2011**, *81*, 1762–1771. [CrossRef]
8. He, F.N.; Li, X.; Zhu, M.K.; Hu, J.H.; Yuan, Y.J.; Li, C.C.; Long, J.J. Color Stripping of Reactive-Dyed Cotton Fabric in a UV/Sodium Hydrosulfite System with a Dipping Manner at Low Temperature. *Cellulose* **2019**, *26*, 4125–4142. [CrossRef]
9. Zhang, Y.; Kang, H.; Hou, H.; Shao, S.; Sun, X.; Qin, C.; Zhang, S. Improved Design for Textile Production Process Based on Life Cycle Assessment. *Clean Technol. Environ. Policy* **2018**, *20*, 1355–1365. [CrossRef]
10. Nieminen, E.; Linke, M.; Tobler, M.; Beke, B.V. EU COST Action 628: Life Cycle Assessment (LCA) of Textile Products, Eco-Efficiency and Definition of Best Available Technology (BAT) of Textile Processing. *J. Clean. Prod.* **2007**, *15*, 1259–1270. [CrossRef]
11. Morita, A.; Ravagnani, M. Life Cycle Assessment in a Textile Process. In Proceedings of the 6th International Workshop Advances in Cleaner Production, São Paulo, Brazil, 24–26 May 2017; pp. 1–6.
12. Kazan, H.; Akgul, D.; Kerc, A. Life Cycle Assessment of Cotton Woven Shirts and Alternative Manufacturing Techniques. *Clean Technol. Environ. Policy* **2020**, *22*, 849–864. [CrossRef]
13. Jacquemin, L.; Pontalier, P.Y.; Sablayrolles, C. Life Cycle Assessment (LCA) Applied to the Process Industry: A Review. *Int. J. Life Cycle Assess.* **2012**, *17*, 1028–1041. [CrossRef]
14. Powar, A.S.; Perwuelz, A.; Behary, N.; Hoang, L.; Aussenac, T. Application of Ozone Treatment for the Decolorization of the Reactive-Dyed Fabrics in a Pilot-Scale Process-Optimization through Response Surface Methodology. *Sustainability* **2020**, *12*, 471. [CrossRef]
15. Bin, A.K.; Roustan, M. Basic Chemical Engineering Concepts for the Design of Ozone Gas-Liquid Reactors. In Proceedings of the 17th World Congress & Exhibition: Ozone and Related Oxidants, Innovative & Current Technologies, Strasbourg, France, 22–25 August 2005; pp. 99–131.
16. Memento Degremont. Available online: https://www.suezwaterhandbook.fr (accessed on 25 August 2020).
17. Benini, L.; Mancini, L.; Sala, S.; Manfredi, S.; Schau, E.M.; Pant, R. *Normalisation Method and Data for Environmental Footprints*; European Institute for Environment and Sustainability: Luxembourg, 2014.
18. Aileni, R.M.; Chiriac, L.; Subtirica, A.; Albici, S.; Dinca, L.C. Aspects of the Hydrophobic Effect Sustainability Obtained in Plasma for Cotton Fabrics. *Ind. Text.* **2019**, *70*, 223–228. [CrossRef]
19. Ma, X.; Wang, X.; Yin, X.; Kan, X.; Wang, Z. Electrochemical Stripping of Cotton Fabrics Dyed with Reactive Black 5 in Water and Wastewater. *Chemosphere* **2018**, *206*, 17–25. [CrossRef] [PubMed]
20. Wencki, K. LCA and LCC of the Ozonation System in Neugut. In Proceedings of the DEMEAU, Dübendorf, Switzerland, 17–18 June 2015.
21. Ortiz, I.M. Life Cycle Assessment as a Tool for Green Chemistry: Application to Kraft Pulp Industrial Wastewater Treatment by Different Advanced Oxidation Processes. Master's Thesis, Univesitat Autònoma de Barcelona, Barcelona, Spain, 2003.
22. Arzate, S.; Pfister, S.; Oberschelp, C.; Sánchez-Pérez, J.A. Environmental Impacts of an Advanced Oxidation Process as Tertiary Treatment in a Wastewater Treatment Plant. *Sci. Total Environ.* **2019**, *694*, 133572. [CrossRef]

Article

Material-Driven Textile Design (MDTD): A Methodology for Designing Circular Material-Driven Fabrication and Finishing Processes in the Materials Science Laboratory

Miriam Ribul [1,*], Kate Goldsworthy [2] and Carole Collet [3]

1. Materials Science Research Centre, Royal College of Art, London SW7 2EU, UK
2. Chelsea College of Arts, University of the Arts London, London SW1P 4JU, UK; k.goldsworthy@chelsea.arts.ac.uk
3. Central Saint Martins, University of the Arts London, London N1C 4AA, UK; c.collet@csm.arts.ac.uk
* Correspondence: miriam.ribul@rca.ac.uk

Abstract: In the context of the circular economy, materials in scientific development present opportunities for material design processes that begin at a raw state, before being introduced into established processes and applications. The common separation of the scientific development of materials from design intervention results in a lack of methodological approaches enabling designers to inform new processes that respond to new material properties. This paper presents the results of a PhD investigation that led to the development and application of a Material-Driven Textile Design (MDTD) methodology for design research based in the materials science laboratory. It also presents the development of the fabrication of a textile composite with regenerated cellulose obtained from waste textiles, resulting from the MDTD methodology informing novel textile processes. The methods and practice which make up this methodology include distinct phases of exploration, translation and activation, and were developed via three design-led research residencies in materials science laboratories in Europe. The MDTD methodology proposes an approach to design research in a scientific setting that is decoupled from a specific product or application in order to lift disciplinary boundaries for the development of circular material-driven fabrication and finishing processes at the intersection of materials science and design.

Keywords: design methodology; materials science; textile recycling; regenerated cellulose; composites; fabrication; material design; transdisciplinary; interdisciplinary; circular economy

Citation: Ribul, M.; Goldsworthy, K.; Collet, C. Material-Driven Textile Design (MDTD): A Methodology for Designing Circular Material-Driven Fabrication and Finishing Processes in the Materials Science Laboratory. *Sustainability* **2021**, *13*, 1268. https://doi.org/10.3390/su13031268

Received: 21 December 2020
Accepted: 20 January 2021
Published: 26 January 2021

Publisher's Note: MDPI stays neutral with regard to jurisdictional claims in published maps and institutional affiliations.

Copyright: © 2021 by the authors. Licensee MDPI, Basel, Switzerland. This article is an open access article distributed under the terms and conditions of the Creative Commons Attribution (CC BY) license (https://creativecommons.org/licenses/by/4.0/).

1. Introduction

A strong focus on the exploration of materials in design and materials science is placed on finding viable alternatives to materials in existing processes and reducing their environmental impacts [1–3]. Scientific advancements are promising factors to enable sustainable change in how we use natural resources [4,5]. These specialist processes, however, are normally removed from a design practice. Technical material developments take place in a scientific context where, according to Küchler, in the nineteenth century, "malleable" materials and new production technologies removed design from the processes of industrial material manufacture [6]. Miodownik describes the start of a complex materials revolution in the twentieth century, where discovery and development became a scientific activity separated from the arts, and argues for a methodological approach in which artists get to know materials through artistic processes [7]. In the context of the complexity of materials science, Manzini was the first to suggest that a material should be described not for what it "is", but for what it is "used for" and to consider, "how does it work" [8] (pp. 55–63). Based on this, Karana et al. ask what a material "expresses to us, what it elicits from us, and what it makes us do" [1] (p. 35). This aligns with Tim Ingold's argument that within the realms of anthropology, art, archaeology and architecture, we need a practice "*with*" and not "*of*" materials [9] (p. 8). He argues against a hylomorphic model where, "practitioners impose forms internal to the mind on a material world 'out there'", but instead concludes that making is a "process of growth" in which materials are "active" participants [10] (pp. 20–21). Attempts to integrate design in scientific material research have led to a new generation of material engagement: materials-by-design [6]. In this approach, the product end-use comes first, and newly developed materials are made to fit technological requirements. The field of circular material innovation would benefit from design-integrated experimentation "with" materials in scientific development in order to develop new materials and processes fit for the circular economy.

There is a methodological gap for design research based in the materials science laboratory that was established as part of the PhD research of this paper's first author, using a literature and a practice review [10]. Acknowledging that the scientific development of materials takes place in dedicated materials science laboratories, we argue that designers can expand materials science research by integrating design research tools at the onset of the material development. The "Krebs Cycle of Creativity", developed by Oxman, places design and science opposite each other in a coordinate plate and connects them through art or engineering [11]. The interactions between these four domains in this cycle evidence exchanges as "currency" and not a methodological approach as such. Whilst a design-science practice was pioneered by Buckminster Fuller in 1927 [12], historically, efforts have been placed on transforming the design method into a scientific one [13]. Karana et al. [1], Peralta [14], Driver et al. [15], and Rust [16] have listed a range of projects that support the collaboration between the disciplines of design and materials science. However, product designers are still determining how to operate in the scientific domain [14,15]. Many design methodologies place ideas and inspiration [17], a vision [18], or design thinking [19,20] at the start of the design process. For example, the Design Council's "double diamond design process model" is represented by areas of divergent and convergent processes in two consecutive "diamond"-shaped stages [17] (p. 6). However, the double diamond methodology is defined by a material selection in the second diamond [21], which is a method also used in engineering [22], in order to apply the material to specific end products or applications. Table 1 summarises the constituent elements of existing material design methodologies in interdisciplinary collaboration: the driver, the methods employed at the start of the research, the setting in which the research takes place, its outcomes, and the mode in which the practice takes place. Examples of material design methodologies in interdisciplinary collaborations are limited. These interdisciplinary projects take place individually in separate domains of the laboratory or design studio [1,23–25], or are facilitated in neutral settings, such as workshops in large-scale projects [26–29]. Moreover, even if designers work from within the materials science laboratory, the practice with materials remains within the disciplinary domain of design or science, and the outputs of these interdisciplinary collaborations are mostly new material developments or applications. A shift from a product or material development focus to concentrating on processes with materials would invert the common design methodology beginning with an envisioned product and application, as it is found in a materials-by-design approach. The context of circularity in which regenerated cellulose materials obtained from waste textiles are chemically recycled is a recent disciplinary domain which would promote such investigation.

Table 1. Drivers, action steps, setting, outcomes, and practice in interdisciplinary material design methodologies.

Author and Year	Title (If Named) and Driver	First Action Step	Setting	Outcomes	Practice
Thong and Jackson, 2011 [30]	Product Design Driven Research	Specification of material performance criteria	Not specified	Commercial application of new materials	Interdisciplinary collaboration
Rognoli et al., 2015 [24]	DIY-materials	Material sampling	Design studio practice	Material samples	New materials for design

Table 1. Cont.

Author and Year	Title (If Named) and Driver	First Action Step	Setting	Outcomes	Practice
Karana et al., 2015 [1]	Material Driven Design (MDD): Materials experience	Technical and experiential characterisation through tinkering	Design studio practice	Material experiences within a material or a product	(Material) design practice
Härkäsalmi et al., 2017 [31]	Technical and perceptual qualities of materials	Design-driven process in a material-based approach defined by an example application	Iterative prototyping in unspecified setting	Application-driven: acoustic interior elements	Interdisciplinary research (design-science collaboration) within the disciplinary domain
Niinimäki et al., 2018 [27]	Properties of new materials	Presentations and assignments (hands-on play with similar materials)	Interdisciplinary workshops	Interdisciplinary materials development: material properties; application areas; design ethics	Interdisciplinary: materials science, synthetic biology, design and art
Barati et al., 2019 [25]	Smart materials	Prototyping material demonstrators	Design studio	Collaborative material development	Interdisciplinary: design, materials science
Tubito et al., 2019 [26]	Design-driven Material Innovation (DDMI): Materials and their technology	Envisioning material and design scenarios	Materials R&D in parallel to material conceptualisation; workshop setting for the collaboration	New applications and systems of materials	Interdisciplinary: design practice, materials science, manufacturing, end user research

Regenerated cellulose obtained from waste textiles has existed since 2012, and only a few design prototypes have been produced from scientific research to demonstrate its potential applications, whilst these interdisciplinary partnerships are difficult to map when they are not documented [10]. Regenerated cellulose is here produced in a non-toxic chemical recycling process in which post-consumer cotton is dissolved before it can be regenerated in a coagulation bath and spun into new fibres [4,32–34]. The scientific developments with regards to regenerated cellulose obtained from waste textiles suggest its potential to replace environmentally impactful cotton fibres [35–38]. Practice-based textile design and materials science collaborations in this field create artefacts to demonstrate the viability of regenerated cellulose to substitute materials in established processes such as knitting, weaving, or 3D printing at the product, finishing, and textile processing stages, from yarn to fabric, in the existing textile value chain, as evidenced in the outputs of these projects [39]. In this approach that aims for a like-for-like replacement of environmentally impactful materials, design research cannot intervene into the scientific development of regenerated cellulose to inform new textile processes. On the other hand, the scientific achievement of being able to regenerate cellulose from end-of-life textiles with non-toxic chemicals results in a material that is suitable for the circular economy. Regenerated cellulose materials sit within the context of the bioeconomy for industries that use biological materials, enabling a circular bioeconomy [40]. Averting the use of landfills for post-consumer textiles through chemical recycling technologies would keep resources within a closed loop. However, scientific research states that cellulose-based materials cannot be infinitely recycled while maintaining the same quality [38,41]. This may provide new challenges for textile processes in the circular economy, since recycled regenerated cellulose materials have a decreased polymer length that would make this material unsuitable for the established textile value chain. This suggests that circularity requires the need for intervening with the raw materials at hand before these are manufactured, processed, or engineered for a specific process or application, as well as textile processes that both enable circularity and respond to the context of circular recycling.

This paper presents a Material-Driven Textile Design (MDTD) methodology for design research based in the materials science laboratory that facilitates design intervention at the first stage of scientific research, in order to develop new circular processes "with" the material. The methodology is the result of a PhD research investigation with the hypothesis that textile design research intervening in the scientific development of regenerated cellulose materials can inform new textile processes inscribed within the circular economy. The design research was structured around three research residencies in materials science laboratories (2016–2018): the first two residencies with Dr Hanna de la Motte, the focus area manager for Circular Materials Ecosystems (AoI Material Transition) and a researcher at the Division of Materials and Production (Department of Chemistry, Biomaterials and Textiles; unit Fiber Development), which was then the Bioeconomy Division (Cellulose-based Textiles Section; Biorefinery Unit) at RISE Research Institutes of Sweden (RISE) [42,43], and the third residency at the Department of Bioproducts and Biosystems of Aalto University's School of Chemical Engineering, in Finland. Each residency corresponds to one of the three action steps described in the methodology, with the first residencies followed by two studio practice stages, structuring the activity into stages of action and reflection [10]. The three stages of the MDTD methodology corresponding to three research residencies in materials science laboratories were developed at the outset of the research. Section 2 outlines the methodological approaches that underpin the development of the methodology. Sections 3–5 then describe the three action steps of exploration, translation and activation, and how the methods in each stage developed through practice, leading to a new circular material-driven process for textile composite fabrication with regenerated cellulose. Section 6 discusses the MDTD methodology in the context of material design methodologies in interdisciplinary research and the challenges of its constituent elements when it is applied by other designers and to other materials. Finally, the conclusion in Section 7 evaluates how design research based in the materials science laboratory can establish new courses of action for the scientific development of new circular and regenerated materials.

2. The Material-Driven Textile Design (MDTD) Methodological Framework

This section describes the methodology developed in a textile design context, but refers to "material design processes", "material design situations" and "material design visions", which include textiles. Figure 1 illustrates the three action steps of exploration, translation, and activation in the author's Material-Driven Textile Design methodology, in which a raw material is the starting point of the research and the results are textile or material artefacts resulting from the new material design processes. Figure 1 also shows which disciplinary domain informed each action step: the material tests in the exploration stage are informed by materials science, represented by a diamond shape; both materials science and design equally inform the material experiments in the translation stage; therefore, the diamond shape merges with an ellipse; the new material design processes in the activation stage are then based on the design vision, represented by an elliptical shape. The designer tests, experiments and designs new material design processes in each action stage by introducing design techniques into the materials science laboratory. These processes result in circular artefacts that are compatible with the raw material stage after recycling. The designer can consequently repeat the methodology to develop processes that respond to the modified material properties.

The Material-Driven Textile Design (MDTD) methodology is underpinned by the theoretical context of three methodological approaches, which evolved from the following principles for a materials design practice situated in the materials science laboratory: action research and participatory design research for the collaboration with materials scientists to access, observe, and participate in scientific processes; material-driven design for the focus of the methodology on exploring new material design processes, which are decoupled from a specific product or application; and tacit knowledge in a strong design disciplinary background to inform the transdisciplinary, practice-based work with materials in the tools and techniques introduced.

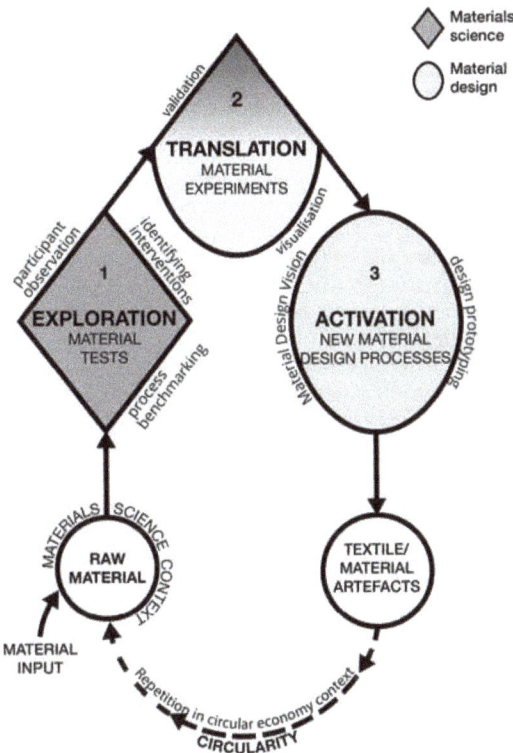

Figure 1. Material-Driven Textile Design (MDTD) methodology [10].

Participatory design in action research can be used to create an equal collaboration with materials scientists. Whereas action research promotes experiments "in the field, rather than laboratory" [44] (p. 18), the field for design research in the MDTD methodology is in fact the materials science laboratory. Action research comes from the social sciences, but instead of research on others, it argues for a critical self-reflection that can take place with others, and therefore focuses on the transformation of practice [45]. The MDTD methodology aims to achieve a change of design practice in a materials science context. This change of practice follows Kurt Lewin's iterative cycles of planning, acting, observing, and reflecting [45]; cycles of action and reflection in constructivist research [46]; and each research cycle can revise the initial plan [47]. Whereas action research can be performed individually, participatory design research evolved with the aim of creating change in society by involving the participants in the research in an equal manner [48–50]. In participatory action research, a cooperative enquiry is a form of research "with" rather than "on" people, where "all the active subjects are fully involved as co-researchers in all research decisions" [49] (p. 145). This methodology can be used for small group research projects, ingraining the transformation of the participants [49]. As the MDTD methodology does not focus on the study of scientists, but on the development of new textile design processes, it can be useful for the development of interdisciplinary collaborative research between individual design and materials science researchers.

The second methodological approach addresses the practice "with" and not "of" materials in their raw state [9] (p. 8), and methods for material-driven design. An approach that puts the material at the start of the design process is "materials driven design", which begins with exploration and experimentation to find new opportunities [21] (p. 282). The "Material Driven Design (MDD) method" of Karana et al. begins with an, "understanding of the material" through "tinkering" in order "to understand its inherent qualities, its constraints, and its opportunities" [1] (p. 41), but differentiates itself from other material-driven approaches by designing for material experiences through a "product and/or further developed material" [1] (p. 10). The mastering of the material through "tinkering" is particularly suitable for materials that are "not fully developed" [1] (p. 41), such as regenerated cellulose obtained from waste textiles. What the MDTD adds to, or replaces in, the MDD method is described through the practice in Sections 3–5.

The third aspect is how tacit knowledge informs the evaluation and progression of the design practice when it is based on craft knowledge such as design techniques. Craft here explores a "flow of activity" [51] (p. 35) and emerges from "embedded knowledge" in "the interplay between tacit knowledge and self-conscious awareness" [52] (p. 50). A recurrent practice with materials leads to tacit knowledge, a specifically intuitive approach that cannot be described in an instruction for others to emulate, but that practitioners can apply to other work. The results of the design practice in the MDTD methodology are analysed with a qualitative assessment of the haptic and visual properties based on tacit knowledge of the disciplinary background. This assessment is formed by actions based on "tacit knowledge", which was first argued by Polanyi to be based on "a rich understanding and knowledge" that is "gained over life time experience, a theory that is increasingly applied to design and artefacts" [16] (p. 77). Tacit knowledge in design is mostly traced back through "reflection-in-action" [46] (p. 49) and becomes evident in the results of processes and techniques such as those introduced into the materials science laboratory in the MDTD methodology, as well as in the manifestation of the "technical", "sensorial", and "aesthetic" character of the resulting materials and artefacts [1] (p. 42).

The next sections describe the methods of the three methodological stages of exploration, translation, and activation in order to illustrate the development of the practice with regenerated cellulose materials in chemical recycling. This research produced two hundred samples, resulting from the experiments in both the materials science laboratory and the studio practice [10]. The following sections document key experiments for each of the action stages towards the development of a new textile composite fabrication process. Whilst experiments were repeated several times for validation, a selected successful sample is included in this paper. Including multiple samples into this paper would hinder an overview of the development of practice through research. The experiments appearing in this paper are numbered according to the corresponding residency, followed by the number of the experiment taking place within each residency.

3. MDTD Action Step 1: Exploration

The exploration stage corresponds to the first residency in the materials science laboratory at RISE Research Institutes of Sweden. Latour and Woolgar argue that scientific processes require an observation "in situ" and claim that from a social science perspective [53] (p. 37), the observer needs to select a "theme" to delineate the method [53] (p. 35), which will produce order from the observations. The theme in the design brief for the first residency was, "to explore the specific properties and processes concerned with the production of regenerated cellulose in the science laboratory" at the raw material stage [10] (p. 365). Language barriers in interdisciplinary research were also considered on site and Wilkes et al. [54] and RISE Research Institutes of Sweden [55], outline that tools for collaboration are often required in these spaces. By employing design methods—such as sketchbooks, drawings (Figure 2a), and mapping (Figure 2b)—to document the way in which both the scientist and design researcher were thinking about textile design research from within the context of working in the materials science laboratory together, these tools helped bridge the discipline-specific language and explore visual communication in the residency [42].

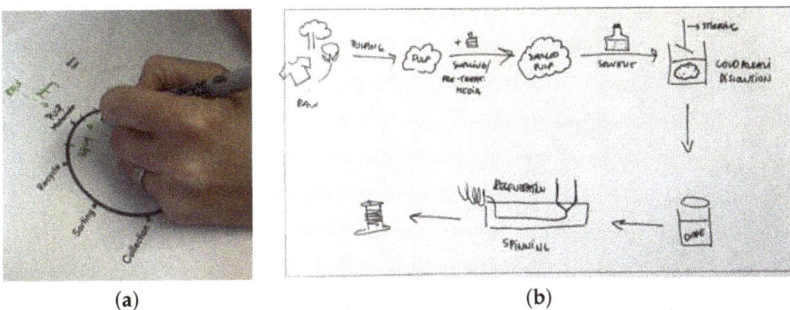

Figure 2. (**a**) Drawing of the circular lifecycle of regenerated cellulose obtained from waste textiles; (**b**) Mapping of the process for cellulose regeneration with the materials scientist.

3.1. Participant Observation

This stage observed the existing scientific method for the dissolution of cellulose materials, the scientific analysis of the cellulose dissolution and its suitability in the spinning process, as well as the scientific method for the regeneration of dissolved cellulose in the fibre spinning process. The observation was documented with a sketchbook, photography, diagrams, and process maps, as well as notes and interviews. One key observation was that when fibres do not dissolve or when the properties of the cellulose dissolution change, the raw material may not be suitable for fibre spinning. This informed two directions for the research: whether new textile processes could make use of this otherwise redundant material, and the lab work follows a scientific method in order to spin a fibre from a cellulose dissolution, in which design research cannot intervene experimentally, as it would disrupt the formation of the fibre. A better understanding of the chemical recycling stage of waste textiles was found when actively performing the scientific processes for the dissolution of cellulose. The design practice at this stage did not deviate from the scientific method that was applied in the laboratory in order to engage with the material properties and the processes as they occur (Figure 3). Participant observation identified the raw material state in which the design practice would intervene: the cellulose dissolution before this is being regenerated into a new form [42].

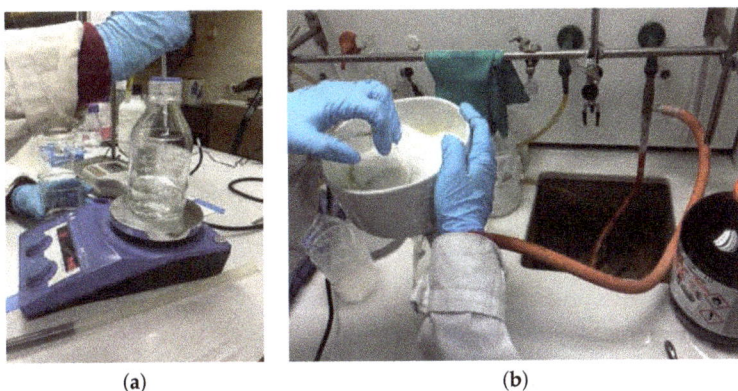

Figure 3. (**a**) Cellulose pre-treatment preparation during the first residency at RISE Research Institutes of Sweden; (**b**) Pressing and dewatering post-consumer cotton for dissolution.

3.2. Mapping Design Interventions

The exploration stage in the first design research residency aimed "to map how design can intervene in the production processes for regenerated cellulose in the scientific laboratory" [10] (p. 365). Interfering with scientific methods involves a high cost due to the people, time, and resources involved [53]. Non-invasive design interventions that do not disrupt the fibre spinning process in scientific research had to be found. Making a film was found to be a suitable process that designers can explore, with a cellulose dissolution that is unsuitable for fibre spinning [43]. Regenerated cellulose films can be produced with shortened cellulosic fibres that are obtained from textiles waste using a cellulose dissolution with a lower degree of polymerisation that cannot be spun into fibres [42]. Researchers may extrude, mould, or dry a regenerated cellulose film following a scientific method, and these are techniques which can be explored through design. Tools and techniques for moulding the films were explored both with regenerated cellulose in the materials science laboratory and with bioplastics with similar properties in the studio practice when access to the laboratory or the material was unavailable due to cost or time constraints [43]. The results were documented with photography, a sketch book, or a "material diary" [56] (p. 131), and through the resulting samples.

3.3. Process Benchmarking

Process benchmarking was developed from "material benchmarking" [1] (p. 41), which places the material in a context of similar materials, their applications, and experiential properties. The benchmarking of processes in the MDTD methodology places regenerated cellulose films into a context of similar cellulose-based materials in order to identify different processes that employ this material and whether textile qualities can be achieved by working with such processes. The literature and practice review of current applications in both science and design found that films in materials science are considered for packaging applications [40], not for textiles, in order to achieve a sustainable replacement of cellophane and its properties [57,58], through extrusion and casting into equal flat shapes [57,59,60]. In textile design, regenerated cellulose film making is not explored outside of the materials science laboratory. Processes are evidenced in printing onto film in packaging applications [33] or in other cellulose-based materials for garment moulding [61], extrusion for architectural structures [62], extrusion of textile yarn [63], and reactive properties of 3D-printed cellulose film shapes [64]. The results of these processes are flexible, transparent films that lack haptic and visual textile properties.

3.4. Practice: Material Tests

Material testing, through participation in the scientific research, facilitated the designer's knowledge of the scientific methods and understanding of how to work effectively with them. Process benchmarking informed the planning of the design tools and techniques for the material experiments. Textile design techniques for moulding were introduced in order to form the film into a range of shapes. The moulds were selected and developed to generate textile structures such as nonwovens and nets (Figure 4a). After producing a series of thin round films, a suitable scientific method for material testing was established. The objective of experiment 1.2, which represents the second experiment in the first residency, was to test this method in a moulding technique. The materials used were a cellulose source of post-consumer cotton provided by the laboratory and the ionic liquid solvent 1-Ethyl-3-methylimidazolium acetate (EmimAc) to dissolve the cotton at 70 °C with a dissolution of 8% cotton in the solvent. The tool introduced for moulding the cellulose dissolution was a flexible aluminium mesh (Figure 4b). The resulting film was regenerated in a water-based coagulation bath and dried in an oven for controlled heat (Figure 4c). The qualitative assessment evidenced that the film bonded to the edges of the metal grid and broke when being removed. The film shrank when dried. It is hard, brittle, transparent and with a texture similar to plastic (Figure 4d). The success criteria showed that the film can be moulded into a fine lace-like shape but evidenced that a different method for regeneration needed testing in order to achieve flexible films. The first experiments identified bonding and moulding design techniques for exploring the cellulose dissolution through design.

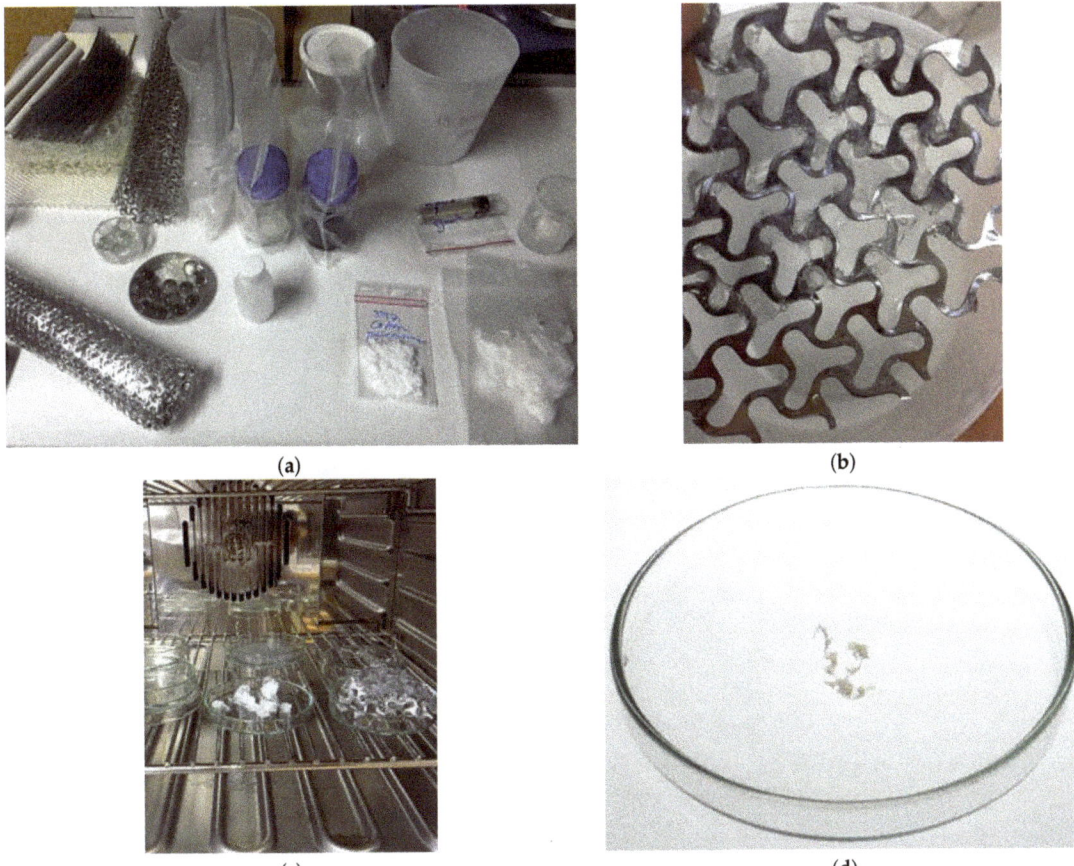

Figure 4. (**a**) Preparation of tools, materials, and moulds for material testing during the first residency at RISE Research Institutes of Sweden; (**b**) Experiment 2.1. Aluminium mesh for moulding the cellulose dissolution; (**c**) Regenerated cellulose samples drying in the laboratory oven; (**d**) Experiment 1.2. Regenerated cellulose lace-shaped film.

4. MDTD Action Step 2: Translation

The translation stage occurred during the second residency at RISE. The translation stage brief outlined how textile design techniques are introduced into the regeneration stages of cellulose, "to explore prototyping with regenerated cellulose films in the science laboratory for a [... new value] chain for textiles from raw material to product" [10] (p. 363). The transdisciplinary methods and the development of the translation stage are further described in Ribul and de la Motte [43]. Both experiments in this stage considered "visualising" the material properties and behaviour at a tangible scale using existing design techniques, as well as the scientific method to "validate" the results with repeatable and shareable processes in the context of the circular economy [43]. The result is a technical material archive that demonstrates the prototyping possibilities with the material and the development of a transdisciplinary material design practice between the two disciplines [43]. The translation stage considered circularity in the material's "past", where the results of the experiments were compatible with the raw material state. Mono-material approaches informed the decisions in the design techniques for the material experiments, as well as the options for disassembly in order to remove any added material at the end of life.

4.1. Visualisation

Before the material experiments began, a wide set of design tools and techniques informed by the textile design disciplinary background was planned in the design studio in order to establish various haptic and visual properties (such as flexibility, texture, or colour) with regenerated cellulose films in textile fabrication and finishing processes. The techniques introduced to visualise the properties of regenerated cellulose included moulding, 3D printing (Figure 5a), bonding, and coating. Each technique demonstrated the limitations of the scientific method in the design techniques: for example, a failed experiment resulted in a material breaking or an extruded dissolution cellulose regenerated on impact in a coagulation bath (Figure 5b).

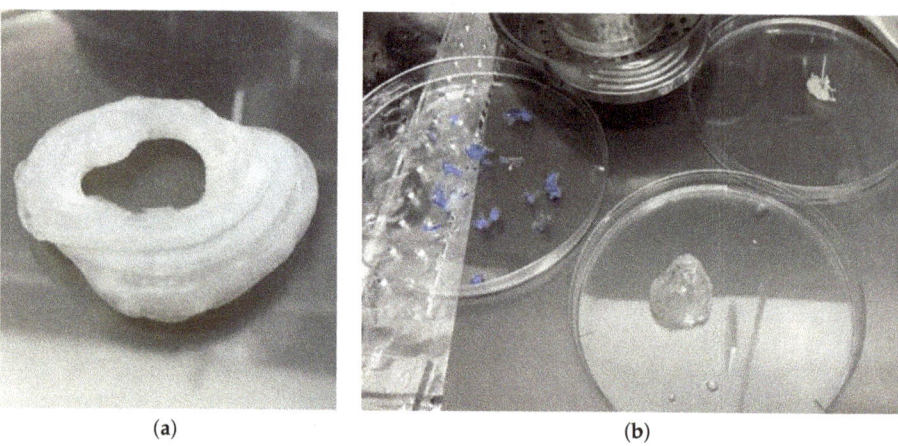

Figure 5. (**a**) Sample showing a three-dimensional extrusion of regenerated cellulose; (**b**) Three samples resulting from experiments including extruded regenerated cellulose that coagulated on impact.

4.2. Validation

The validation of the material experiments was informed by materials science in the following stages: (1) by adopting the scientific method in the design techniques, and (2) by documenting the experiments with a lab book. It was pertinent to the translation stage that material experiments occurred in the materials science laboratory and that they used the material that is the focus of the research. The validation of the experiments with regenerated cellulose repeated the scientific method by introducing variables such as different waste textiles, solvents, and settings in order to find the most suitable one for applying textile techniques [43]. A method was found that creates flexible films, which was then also utilised in residency 3. The method showed which design techniques can be introduced at the raw material stage and identified where techniques should be discarded or adapted.

4.3. Practice: Material Experiments

The material experiments followed a similar approach to the "exploration" [21], or the first step of the "Material Driven Design (MDD)" method in the technical characterisation of the material by tinkering [1]. In the MDD method, the technical stage is followed by focus groups and interviews to map the "experiential characterization" [1] (p. 41) that the material may elicit in products. This was not the case in the translation stage, where the material experiments progressed towards textile processes with a "knowing-in-action" and "reflecting-in-action" approach [46] (p. 49) to evaluate the "sensoaesthetic" [7] (p. 69) properties of the results.

The objective of experiment 2.13, which represents the 13th experiment in the second residency, was to mould the cellulose dissolution into a three-dimensional form. An undyed 100% post-consumer cotton textile provided by the laboratory was dissolved using the ionic liquid 1-Ethyl-3-methylimidazolium acetate (EmimAc) at 80 °C with a dissolution of 5% cotton in the solvent. An additive of sawdust was introduced in order to create texture and colour, and the tool used in the design technique was a three-dimensional plastic mould (Figure 6a). The dissolution was then regenerated in an ethanol coagulation bath and air-dried on a metal mesh. The qualitative assessment evidenced that the composite shrank less when drying and kept the shape of the mould, while the sample looks like wood and feels like paper (Figure 6b). This result informed further testing of moulded films with additive particles in order to reduce shrinking in three-dimensional mono-material composites. The translation stage identified four mono-material design techniques for the circularity of the material: colour, texture, print, and form [10].

Figure 6. (**a**) Experiment 2.13. Plastic mould for 3D moulding; (**b**) Experiment 2.13. 3D-moulded film with sawdust.

5. MDTD Action Step 3: Activation

The activation stage is the final action step in the MDTD methodology. Here, the findings from the exploration and translation stages informed the Material Design Visions for new textile processes using regenerated cellulose materials obtained from waste textiles. An iterative cycle of design prototyping took place in the third residency at Aalto University's School of Chemical Engineering, in order to develop tangible textile or material artefacts that act "as an embodiment for a hypothesis, realizing the conditions (independent variables) in an experiment" [65] (p. 95), and materialise a possible future with a material design process that can be evaluated against textiles resulting from existing processes. The aim of residency 3 was, "to create a range of textile samples and artefacts that emerge from a practice with regenerated cellulose materials at the intersection of design and science that demonstrate [. . . new] textile processes for the circular bioeconomy" [10] (p. 367). The literature review and process benchmarking in the exploration stage established whether new processes with the material have been achieved. In this final stage, a "future" circularity perspective informs design processes that consider working with materials with modified properties after mechanical or chemical recycling has taken place.

5.1. Material Design Visions

According to Verganti, "envisioning" occurs when a designer creates new meanings with their design [18] (p. 180). In the MDD method by Karana et al. [1] (pp. 42–43), a "Material Experience Vision" is the second action step after the technical and experiential material characterisations have been completed. The properties of chemically recycled cellulose in existing design techniques in the translation stage informed the development of Material Design Visions for processes that manifest new forms of "material design" in a transdisciplinary domain. The envisioned techniques synthesised the results from the previous two action steps: (1) the properties of the material in the exploration stage, and (2) the circular design techniques in the translation stage. Four Material Design Visions were formed for processes and haptic and visual properties that are distinguished from existing developments in this field: textile shape and surface manipulation; nonwoven textile fabrication; colour in the finishing process; and 3D-moulded composites. Prototyping tools were prepared in the design studio practice in response to the envisioned textile techniques.

5.2. Design Prototyping

If the aim of design is "creating something that does not yet exist (either knowledge or product) and that fits into the future", then prototypes help to visualise this new paradigm and to communicate it in a tangible way [65] (p. 85). Prototypes in research can validate an idea and are often used for "testing a theory" or a "hypothesis" [65] (p. 95), but can also play a role in "reflecting on open-ended exploration" [65] (p. 87). The aim of the prototyping in the activation stage was to develop new textile fabrication and finishing processes with regenerated cellulose. The prototyping process in itself generated concrete information about the design to optimise the envisioned textile techniques and to establish the desired material outcomes. The results of the experiments were therefore qualitatively evaluated against the haptic and visual properties of textiles such as texture, form, strength, lightness, drape, colour, composition, and thickness.

5.3. Practice: New Material Design Processes

The final set of experiments produced artefacts that demonstrate the change in the textile processes obtained through the action research carried out in the materials science laboratory at RISE. The experiments followed the scientific method described in Section 4.3, except for the use of the Ioncell solvent, which was developed and patented at Aalto University [66]. Prototyping either proved or disproved the hypothesis of the Material Design Visions and refined the techniques employed for creating the final artefacts. For example, the envisioned 3D-moulded composite technique was discarded due to the fact that the experiments from the translation stage could not be validated using the Ioncell solvent. Prototyping, in turn, achieved a new process to fabricate a flexible textile composite. The composite fabrication process in experiment 3.26, which represents the 26th experiment in the third residency, used the same 100% post-consumer cotton waste used in experiment 2.13 and described in Section 4.3, the Ioncell ionic liquid to dissolve it, and an additive of recycled black cotton fabric. The objective of the experiment was to form a composite that has haptic and visual textile properties such as drape, handle, lightness, and breathability. A modified textile printing technique was developed to deposit the cellulose dissolution [67], which was regenerated in a water-based coagulation bath, and the result was dried in a humidity-controlled cabinet for 54.15 h, with humidity ranging from 50% for 20.15 h to 25% for 29 h and 10% for 5 h. The assessment of qualitative properties revealed a cellulose-based textile composite that feels soft and light and can be draped, breaking slightly at the edges (Figure 7). The success criteria evidenced the impact of the drying method on the outcome, informing further experiments. The practice-based work with the cellulose dissolution in the materials science laboratory established four new processes as the result of the methodology, which comprise two fabrication and two finishing processes [10,67]. Each one of these processes evolved in parallel to the others, in response to the three residencies starting from the exploration of material tests, followed by the translation of material experiments, and finally the activation of new processes through prototyping.

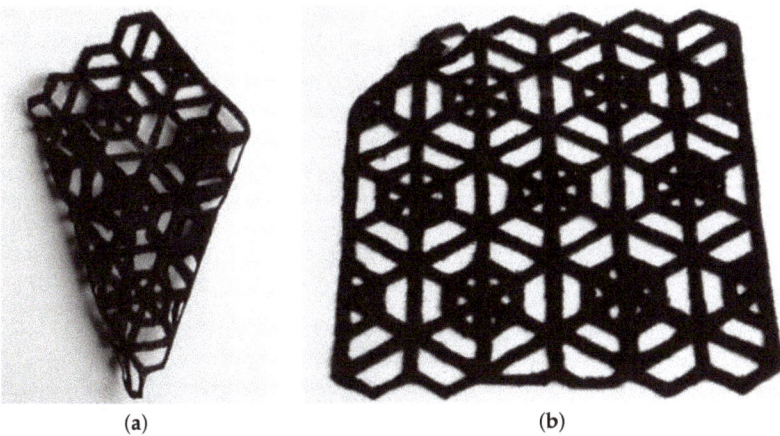

Figure 7. (a) Experiment 3.26. Textile composite folded onto itself; (b) Experiment 3.26. Textile composite.

6. Discussion

This paper has presented the author's Material-Driven Textile Design (MDTD) methodology, which enables designers to test, experiment and design new material design processes at the "raw" stage of their scientific development. This methodology was developed and applied with an investigation of regenerated cellulose obtained from waste textiles in the context of the circular economy. As opposed to making a "material selection" for specific end products [21], the design practice presented in this paper starts from an exploration "with" the material, proceeds with a "translation" of the design-science practice and results in an "activation" of new material design processes. The MDTD methodology offers the opportunity to go beyond an "exploration" of materials close to their raw form. Moreover, it offers an effective practice-based approach leading to new material design processes that are scientifically validated and where the haptic and visual properties of the results can be evaluated against existing processes in textile design.

Table 2 evaluates constituent elements of the MDTD methodology in the context of interdisciplinary research projects (described in Table 1). The MDTD methodology distinguishes itself from existing material design methodologies in that it integrates design practice into scientific development and the scientific method into design practice. The methodology lifts the boundaries of separated disciplinary domains, in which materials science and design usually operate, and establishes a setting in which design participates in scientific research for material development "in situ", enabling a new transdisciplinary practice to emerge. Primarily, the MDTD methodology creates a new methodological framework in which the material at hand and its properties at their "raw" stage are the drivers that inform new material design processes, decoupled from envisioned products or applications in the prevalent hylomorphic model of design. In this context, the designer's and the scientist's role does not aim for material or product development, but instead, the practice "with" the material results in new processes that inform new models for fabrication and finishing. The context of a circular economy requires a reframing of the common model of like-for-like replacement in scientific material development, in order to make new, regenerative approaches possible.

Table 2. Drivers, action steps, setting, outcomes, and practice in the MDTD methodology.

Author and Year	Title (If Named) and Driver	First Action Step	Setting	Outcomes	Practice
Ribul, 2019 [10]	Material-Driven Textile Design (MDTD): Material in its raw state and its properties	Exploration: participant observation, mapping design interventions, and process benchmarking	Based within the materials science laboratory	New material design processes "with" materials at their raw stage	Transdisciplinary, fluid integration of design and materials science practice

The MDTD methodology resulted in new textile processes inscribed within the circular economy, including the process for textile composite fabrication presented in this paper. Circular and regenerated materials can have properties that make them unsuitable for established processes and applications. The application of the MDTD methodology to a material in scientific development different from the one described in this paper will have different limitations and opportunities for design to intervene and develop new processes. Each "raw" material will require an in-depth exploration before new processes can be activated. These materials could be related to the disciplinary practice of designers or present unexplored starting points.

Challenges arising from the application of the methodology by other designers may be three-fold. The first challenge is that the experiments "with" materials in the three action stages depend on the tacit knowledge from a broad range of techniques of the designer, resulting from previous design projects. The second challenge is to avoid defining products or applications "of" the material early on, which could be explored only after the activation stage is completed. The third challenge is that designers may limit themselves to the exploration stage and focus solely on material tests. Without the translation stage, the designer may not identify a range of design techniques and develop a transdisciplinary practice. Similarly, the exploration and translation stages alone would hinder the development of new design processes.

The materials science context in the MDTD methodology is imperative. A designer can mimic scientific processes, but tools or materials may not be available, leading to speculative outcomes. Embedding the design practice into a materials science laboratory is recommended even if in the form of short visits and testing, whilst complementing the research with studio practice to anchor the new processes and results in the design disciplinary domain. The challenges here can lie in designers establishing collaborations with materials scientists, access, time, and costs in order to develop new material design processes, which may lead to the adaptation of the three action stages, as well as to new methods applied by each designer in order to achieve results. On the other hand, the designer's disciplinary background and tacit knowledge may adapt and change the methods and stages in this methodology.

7. Conclusions

The methodology described in this paper makes a compelling argument for designers to be active in the materials science laboratory in order to establish new circular material-driven fabrication and finishing processes. Having created and applied the methodology in the context of an investigation of regenerated cellulose and its changing properties in circularity over an extended period of time, the research led to a transformation of the practice in interdisciplinary design and materials science collaboration into one that integrates discipline-specific methods. The methodology was structured into three action stages—exploration, translation, and activation corresponding to three research residencies, each with its own set of methods. The significance of these three action research stages lies in their enabling of new courses of action for materials originating in the materials science laboratory beyond established textile processes and applications. The cellulose-based composite revealed a new textile fabrication process with regenerated cellulose in a circular lifecycle, that is mono-material and compatible with the raw material stage while achieving textile haptic and visual properties. The possible context of use of this textile composite and its potential applications could form a future research project stemming from this research using another methodology. The MDTD methodology and its development are described in this paper in order to support designers who wish to move into a scientific domain whilst retaining their core design knowledge. Its application by other designers would enable a transdisciplinary practice for working "with" materials in their raw state and enable design for circularity in future textile recycling contexts.

Author Contributions: Conceptualisation, M.R.; methodology, M.R.; validation, M.R.; formal analysis, M.R.; investigation, M.R.; resources, M.R.; data curation, M.R.; writing—original draft preparation, M.R.; writing—review and editing, M.R., K.G., and C.C.; visualisation, M.R.; supervision, K.G. and C.C.; funding acquisition, M.R. and K.G. All authors have read and agreed to the published version of the manuscript.

Funding: The PhD research was funded by the AHRC London Doctoral Design Centre (LDoC) at the University of the Arts London (UAL).

Institutional Review Board Statement: The study was approved by the Ethics Committee of the University of the Arts London (27 January 2016).

Informed Consent Statement: Informed consent was obtained from all subjects involved in the study.

Data Availability Statement: The data presented in this study are available on request from the corresponding author. The data are not publicly available due to the embargo on the Ph.D. Thesis of the author at the time of publication of this article.

Acknowledgments: With thanks to the Centre for Circular Design at Chelsea College of Arts (UAL), where this PhD research was based, to Hanna de la Motte and to RISE Research Institutes of Sweden, for hosting two research residencies, and Aalto University for supporting the third residency with laboratory space and materials for prototyping.

Conflicts of Interest: The authors declare no conflict of interest.

References

1. Karana, E.; Barati, B.; Rognoli, V.; Zeeuw van der Laan, A. Material driven design (MDD): A method to design for material experiences. *Int. J. Des.* **2015**, *9*, 35–54.
2. Boston Consulting Group. Global Fashion Agenda. The Pulse of the Fashion Industry. Available online: https://www.copenhagenfashionsummit.com/wp-content/uploads/2017/05/Pulse-of-the-Fashion-Industry_2017.pdf (accessed on 19 May 2017).
3. Ellen MacArthur Foundation. The New Plastics Economy: Catalyzing Action. Available online: https://www.ellenmacarthurfoundation.org/publications/new-plastics-economy-catalysing-action (accessed on 14 March 2017).
4. Kääriäinen, P.; Tervinen, L. (Eds.) *Lost in the Wood(s): The New Biomateriality in Finland*, 1st ed.; Aalto University: Helsinki, Finland, 2017.
5. HM Government. Eight Great Technologies: Infographics. Available online: https://www.gov.uk/government/uploads/system/uploads/attachment_data/file/249255/eight_great_technologies_overall_infographic.pdf (accessed on 23 October 2017).
6. Küchler, S. Materials: The story of use. In *The Social Life of Material: Studies in Materials and Society*, 1st ed.; Drazin, A., Küchler, S., Eds.; Bloomsbury Publishing: London, UK, 2015; pp. 267–282.
7. Miodownik, M. Toward designing new sensoaethetic materials: The role of material libraries. In *The Social Life of Material: Studies in Materials and Society*, 1st ed.; Drazin, A., Küchler, S., Eds.; Bloomsbury Publishing: London, UK, 2015; pp. 69–79.
8. Manzini, E.; Cau, P. *The Material of Invention*, 1st ed.; Arcadia: Milano, Italy, 1986.
9. Ingold, T. *Making: Anthropology, Archaeology, Art and Architecture*; Routledge: Abingdon, UK, 2013.
10. Ribul, M. Material Driven Textile Design: Designing Fully Integrated Fabrication and Finishing Processes with Regenerated Cellulose in the Materials Science Laboratory. Ph.D. Thesis, University of the Arts London, London, UK, 2019.
11. Oxman, N. Age of Entanglement. *J. Des. Sci.* **2016**. Available online: http://www.pubpub.org/pub/AgeOfEntanglement (accessed on 9 June 2016). [CrossRef]
12. Brown, H.; Cook, R.; Gabel, M. Environmental Design Science Primer. Available online: https://bfi.org/sites/default/files/attachments/pages/EnvDesignPrimer-BrownCookGabel.pdf (accessed on 9 June 2016).
13. Cross, N. Designerly Ways of Knowing: Design Discipline Versus Design Science. *Des. Issues* **2001**, *17*, 49–55. [CrossRef]
14. Peralta, C. Collaboration between Designers and Scientists in the Context of Scientific Research. Ph.D. Thesis, University of Cambridge, Cambridge, UK, 2013.
15. Driver, A.; Peralta, C.; Moultrie, J. Exploring How Industrial Designers Can Contribute to Scientific Research. *Int. J. Des.* **2011**, *5*, 17–28.
16. Rust, C. Design Inquiry: Tacit Knowledge and Invention in Science. *Des. Issues* **2004**, *20*, 76–85. [CrossRef]
17. Design Council. A study of the Design Process. Available online: http://www.designcouncil.org.uk/sites/default/files/asset/document/ElevenLessons_Design_Council%20(2).pdf (accessed on 19 July 2016).
18. Verganti, R. *Design-Driven Innovation: Changing the Rules of Competition by Radically Innovating What Things Mean*; Harvard Business Press: Boston, MA, USA, 2009.
19. Lloyd, P. You make it and you try it out: Seeds of design discipline futures. *Des. Stud.* **2019**, *65*, 167–181. [CrossRef]
20. IDEO.org. Design Kit. Available online: http://www.designkit.org/methods (accessed on 10 November 2015).

21. Van Bezooyen, A. Materials Driven Design. In *Materials Experience: Fundamentals of Materials and Design*, 1st ed.; Karana, E., Pedgley, O., Rognoli, V., Eds.; Butterworth-Heinemann: Amsterdam, The Netherlands, 2014; pp. 277–286.
22. Ashby, M.; Johnson, K. *Materials and Design: The Art and Science of Material Selection in Product Design*, 1st ed; Butterworth-Heinemann: Oxford, UK, 2010.
23. Rognoli, V.; Bianchini, M.; Maffei, S.; Karana, E. DIY materials. *Mater. Des.* **2015**, *86*, 692–702. [CrossRef]
24. Rognoli, V.; Garcia, C.A.; Parisi, S. The material experiences as DIY materials: Self-production of wool filled starch based composite (NeWool). *Mak. Futures J.* **2016**, *4*, 1–9.
25. Barati, B.; Karana, E.; Hekkert, P. Prototyping materials experience: Towards a shared understanding of underdeveloped smart material composites. *Int. J. Des.* **2019**, *13*, 21–38.
26. Tubito, C.; Earley, R.; Goldsworthy, K.; Hornbuckle, R.; Niinimäki, K.; Östmark, E.; Sarbach, V.; Tanttu, M. Applied DDMI: A White Paper on How Design-Driven Material Innovation Methodology Was Applied in the Trash-2-Cash Project. Available online: https://static1.squarespace.com/static/5891ce37d2b857f0c58457c1/t/5c74266be2c4830ed71fb71e/1551115936465/D_1_7-White+Paper-MCI-T2C.pdf (accessed on 4 May 2019).
27. Niinimäki, K.; Groth, C.; Kääriäinen, P. New silk: Studying experimental touchpoints between material science, synthetic biology, design and art. *Temes Disseny* **2018**, *34*, 34–43. [CrossRef]
28. Wennberg, M.V.; Östlund, Å. MISTRA Future Fashion Final Report. Available online: http://mistrafuturefashion.com/wp-content/uploads/2019/10/the-Outlook-Report_Mistra-Future-Fashion-Final-Program-Report_31-okt-2019.pdf (accessed on 10 December 2019).
29. Dell'Era, C.; Magistretti, S.; van Rijn, M.; Tempelman, E.; Verganti, R.; Öberg, Å. The White Book: Lessons from a Four-Year Journey into Design-Driven Materials Innovation. Available online: http://elearning.instituteofmaking.org.uk/uploads/dell-era-et-al-2016.pdf (accessed on 13 October 2016).
30. Thong, C.; Jackson, S. Microwave Modified Timber: Collaborative Research integrating Product Design for New Materials Development. *Adv. Mater. Res.* **2011**, *284*, 615–619. [CrossRef]
31. Härkäsalmi, T.; Lehmonen, J.; Itälä, J.; Peralta, C.; Siljander, S.; Ketoja, J. Design-driven integrated development of technical and perceptual qualities in foam-formed cellulose fibre materials. *Cellulose* **2017**, *24*, 5053–5068. [CrossRef]
32. Kataja, K.; Kääriäinen, P. (Eds.) Designing Cellulose for the Future. Available online: https://cellulosefromfinland.fi/wp-content/uploads/2018/09/DWoC_Loppuraportti_FINAL_s%C3%A4hk%C3%B6inen.pdf (accessed on 21 November 2018).
33. Aalto University; VTT; Tampere University of Technology. Design Driven Value Chains in the World of Cellulose DWoC 2013–2015. Available online: http://www.vtt.fi/Documents/DWoC1.pdf (accessed on 28 February 2017).
34. Design Meets Cellulose. CHEMARTS 2020. Available online: http://chemarts.aalto.fi/wp-content/uploads/2017/08/CHEMARTS12_Collaboration.pdf (accessed on 5 October 2017).
35. Ma, Y.; Hummel, M.; Määttänen, M.; Sixta, H. Upcycling of Waste Paper and Cardboard to Textiles. *Green Chem.* **2015**, *18*, 858–866. [CrossRef]
36. Ma, Y.; Asaadi, S.; Johansson, L.; Ahvenainen, P.; Reza, M.; Alekhina, M.; Rautkari, L.; Michud, A.; Hauru, L.; Hummel, M.; et al. High-Strength Composite Fibers from Cellulose–Lignin Blends Regenerated from Ionic Liquid Solution. *ChemSusChem* **2015**, *8*, 4030–4039. [CrossRef]
37. Östlund, Å.; Wedin, H.; Bolin, L.; Berlin, J.; Jönsson, C.; Posner, S.; Smuk, L.; Eriksson, M.; Sandin, G. Textilåtervinning: Tekniska Möjliheter och Utmaningar. Available online: https://www.naturvardsverket.se/Documents/publikationer6400/978-91-620-6685-7.pdf?pid=15536 (accessed on 5 July 2016).
38. MISTRA Future Fashion. Future Fashion Manifesto. Available online: http://mistrafuturefashion.com/wp-content/uploads/2017/12/Future-Fashion-Manifesto-2015-1.pdf (accessed on 15 December 2015).
39. Hummel, M.; Michud, A.; Tanttu, M.; Netti, E.; Asaadi, S.; Ma, Y.; Sixta, H. High Strength Fibers from Various Ligno-Cellulosic Materials Using the Ioncell-F Technology. Available online: http://costfp1205.com/wp-content/uploads/2017/schools/Documents/3rdtrainingschool/2_Hummel_HighStrength.pdf (accessed on 20 June 2016).
40. Ellen MacArthur Foundation. Urban Biocycles. Available online: https://www.ellenmacarthurfoundation.org/assets/downloads/publications/Urban-Biocycles_EllenMacArthurFoundation_21-06-2017.pdf (accessed on 16 September 2017).
41. Lu, J.; Hamouda, H. Current status of Fiber Waste Recycling and Its Future. *Adv. Mater. Res.* **2014**, *878*, 122–131. [CrossRef]
42. Ribul, M.; de la Motte, H. The Material Affinity of Design and Science for a Circular Economy. In Proceedings of the Circular Transitions Conference, London, UK, 23–24 November 2016; pp. 236–248. Available online: http://circulartransitions.org/media/downloads/Circular-Transitions-Proceedings.pdf (accessed on 15 January 2018).
43. Ribul, M.; de la Motte, H. Material Translation: Validation and Visualization and Validation as Transdisciplinary Methods for Textile Design and Materials Science in the Circular Bioeconomy. *JTDRP* **2018**, *6*, 66–88. [CrossRef]
44. Gustavson, B. Theory and Practice: Mediating the Discourse. In *Handbook of Action Research*; Reason, P., Bradbury, H., Eds.; SAGE: London, UK, 2006; pp. 17–26.
45. McNiff, J. *Action Research: Principles and Practice*, 3rd ed.; Routledge: London, UK, 2013.
46. Schön, D.A. *The Reflective Practitioner: How Professionals Think in Action*, 3rd ed.; Ashgate: Aldershot, UK, 1991.
47. Costello, P.J.M. *Action Research*, 1st ed.; Continuum: London, UK, 2003.
48. Simonsen, J.; Robertson, T. (Eds.) *Routledge International Handbook of Participatory Design*, 1st ed.; Routledge: New York, NY, USA, 2013.

49. Heron, J.; Reason, P. The Practice of Co-operative Enquiry: Research 'with' rather than 'on' People. In *Handbook of Action Research*, 2nd ed.; Reason, P., Bradbury, H., Eds.; SAGE: London, UK, 2006; pp. 144–154.
50. Villari, B. Action Research Approach in Design Research. In *The Routledge Companion to Design Research*, 1st ed.; Rodgers, P.A., Yee, J., Eds.; Routledge: New York, NY, USA, 2015; pp. 306–316.
51. Malafouris, L. At the Potter's Wheel: An Argument for Material Agency. In *Material Agency: Towards a Non-Anthropocentric Approach*, 1st ed.; Knappett, C., Malafouris, L., Eds.; Springer: New York, NY, USA; London, UK, 2008; pp. 19–36.
52. Sennett, R. *The Craftsman*, 1st ed.; Allen Lane: London, UK, 2008.
53. Latour, B.; Woolgar, S. *Laboratory Life: The Construction of Scientific Facts*, 1st ed.; Princeton University Press: Princeton, NJ, USA, 1986.
54. Wilkes, S.; Wongsriruksa, S.; Howes, P.; Gamester, R.; Witchel, H.; Conreen, M.; Laughlin, Z.; Miodownik, M. Design tools for interdisciplinary translation of material experiences. *Mater. Des.* **2015**, *90*, 1228–1237. [CrossRef]
55. RISE Research Institutes of Sweden. Third Milestone Report. Available online: https://static1.squarespace.com/static/5891ce3 7d2b857f0c58457c1/t/5c0554e61ae6cfe5a110c824/1543853290648/D9.4+Third+Milestone+Report.pdf (accessed on 7 December 2018).
56. Karana, E.; Blauwhoff, D.; Hultink, E.-J.; Camere, S. When the material grows: A case study on designing (with) mycelium-based materials. *Int. J. Des.* **2018**, *12*, 119–136.
57. Wawro, D.; Hummel, M.; Michud, A.; Sixta, H. Strong Cellulosic Film Cast from Ionic Liquid Solutions. *Fibres Text. East. Eur.* **2014**, *22*, 35–42.
58. Yang, Q.; Fukuzumi, H.; Saito, T.; Isogai, A.; Zhang, L. Transparent Cellulose Films with High Gas Barrier Properties Fabricated from Aqueous Alkali/Urea Solutions. *Biomacromolecules* **2011**, *12*, 2766–2771. [CrossRef]
59. Sundberg, J.; Toriz, G.; Gatenholm, P. Moisture induced plasticity of amorphous cellulose films from ionic liquid. *Polymer* **2013**, *54*, 6555–6560. [CrossRef]
60. Hameed, N.; Guo, Q. Blend films of natural wool and cellulose prepared from an ionic liquid. *Cellulose* **2010**, *17*, 803–813. [CrossRef]
61. Designboom. Suzanne Lee: Eco Textile Fashion. Available online: https://www.designboom.com/design/suzanne-lee-eco-textile-fashion/ (accessed on 15 March 2017).
62. Duró, J.; Mogas, L. Dumo Lab. Available online: http://dumolab.com/ (accessed on 7 March 2017).
63. AlgiKnit. Biology Is the Future of Fashion. Available online: https://www.algiknit.com/ (accessed on 7 March 2017).
64. Wyss Institute. 4D Printing of Shapeshifting Devices. Available online: https://wyss.harvard.edu/technology/4d-printing/ (accessed on 3 January 2017).
65. Stappers, S. Prototypes as a central vein for knowledge development. In *Prototype: Design and Craft in the 21st Century*, 1st ed.; Valentine, L., Ed.; Bloomsbury: London, UK, 2013; pp. 85–97.
66. Michud, A.; King, A.W.T.; Parviainen, A.P.; Sixta, H.; Hauru, L.; Hummel, M.; Kipeläinen, I.A. Process for the Production of Shaped Cellulose Articles from a Solution Containing Pulp Dissolved in Disttilable Ionic Liquids. Patent No. 2014/162062 A1, 4 April 2014.
67. Ribul, M. Future Materials Circularity: The lifecycles of regenerated cellulose in circular fabrication and finishing processes. 2021; manuscript in preparation.

Article

Life Cycle Based Comparison of Textile Ecolabels

Felice Diekel *, Natalia Mikosch *, Vanessa Bach and Matthias Finkbeiner

Department of Sustainable Engineering, Institute of Environmental Technology, Technische Universität Berlin, 10623 Berlin, Germany; vanessa.bach@tu-berlin.de (V.B.); matthias.finkbeiner@tu-berlin.de (M.F.)
* Correspondence: felice.diekel@gmail.com (F.D.); natalia.mikosch@tu-berlin.de (N.M.); Tel.: +49-157-3071-2129 (F.D.); +49-30-314-28455 (N.M.)

Abstract: Environmental impacts of textile production increased over the last decades. This also led to an increasing demand for sustainable textiles and ecolabels, which intend to provide information on environmental aspects of textiles for the consumer. The goal of the paper is to assess selected labels with regard to their strengths and weaknesses, as well as their coverage of relevant environmental aspects over the life cycle of textiles. We applied a characterization scheme to analyse seven selected labels (Blue Angel Textiles, bluesign®, Cotton made in Africa (CMiA), Cradle to Cradle Certified^{TM}, Global Organic Textile Standard (GOTS), Global Recycled Standard (GRS), VAUDE Green Shape), and compared their focus to the environmental hotpots identified in the product environmental footprint case study of t-shirts. Most labels focus on the environmental aspects toxicity, water use, and air emissions predominantly in the upstream life cycle phases of textiles (mainly garment production), whereas some relevant impacts and life cycle phases like water in textile use phase remain neglected. We found significant differences between the ecolabels, and none of them cover all relevant aspects and impacts over the life cycle. Consumers need to be aware of these limitations when making purchase decisions.

Keywords: textile life cycle; environmental aspects; ecolabel; sustainable textiles

Citation: Diekel, F.; Mikosch, N.; Bach, V.; Finkbeiner, M. Life Cycle Based Comparison of Textile Ecolabels. *Sustainability* **2021**, *13*, 1751. https://doi.org/10.3390/su13041751

Academic Editors: Hanna de la Motte and Asa Ostlund

Received: 18 December 2020
Accepted: 3 February 2021
Published: 6 February 2021

Publisher's Note: MDPI stays neutral with regard to jurisdictional claims in published maps and institutional affiliations.

Copyright: © 2021 by the authors. Licensee MDPI, Basel, Switzerland. This article is an open access article distributed under the terms and conditions of the Creative Commons Attribution (CC BY) license (https://creativecommons.org/licenses/by/4.0/).

1. Introduction

The urgency of the climate crisis is more present now than ever before, with the "International Panel on Climate Changes" (IPCC) special report on global warming [1], more than 11,000 scientists warning of a climate emergency [2], and millions of people on the streets for the largest climate strike ever seen [3–6]. The scientists describe a close link between the excessive consumption of a wealthy lifestyle and the climate crisis, naming the global north as mainly responsible for the historic and current greenhouse gas (GHG) emissions [2]. One industry with a particularly devastating impact on the environment is the fashion industry. Apart from a vast contribution to the climate change (in 2015, the textile production alone was responsible for around 1.2 billion CO_2 equivalents of GHG emissions [7]), it is responsible for a whole host of environmental impacts occurring in different life cycle stages of textile products. These impacts include overuse of water resources and excessive use of pesticides during cotton cultivation, contamination of water bodies with untreated wastewater discharged from the textile processing, or pollution with microplastics during the use phase [8]. From 2000 to 2015, the production of clothing has doubled [7]. Due to this constant growth of the fashion industry [9], the environmental impacts associated with textile production are also steadily increasing. This effect was multiplied by a shift in the fashion industry in 1990 towards a fast fashion concept, which lead to an uptake in the speed of production and buying cycles.

At the same time, during the past decades, the awareness of the environmental issues associated with the textile production has continuously increased. A recent study demonstrates that 72% of consumers worldwide would prefer to buy from environmentally friendly brands [10]. As a result, during the last 40 years, various organizations and initiatives emerged using sustainability standards, labels, audits, certificates, or management strategies to enforce sustainable value creation generally referred to as environmental labels and information schemes (ELIS) [11,12]. These include ecolabelling (e.g., Global Organic Textile Standard, GOTS [13]), umbrella ecolabels (e.g., Grüner Knopf [14]), and initiatives for sustainable cotton production (Better Cotton Initiative, BCI [15]) and textiles (Zero Discharge of Hazardous Chemicals, ZDHC [16]). These initiatives aim to make sustainability assessments of textile products easier and provide guidance for consumers. However, as the various labels follow different approaches and have different focus, it remains difficult for consumers to identify relevance and quality of the information the labels offer.

Recently, several studies were conducted with the aim to review and compare textile ecolabels. While some publications provide an evaluation of a single label, e.g., Blue Angel [17] and C2C Certified [18,19], others analyse similarities and differences between the scope and criteria of different labels. For example, Koszewska provides an overview of most popular textile ecolabel and their recognisability among Polish buyers [20]. Existing comparisons between textile ecolabels consider different environmental aspects and life cycle stages of textiles. Partzsch et al. analyse the effect of the certification on cotton cultivation in Sub-Saharan Africa with regard to the use of fertilizers, pesticides, and genetically modified organisms (GMO) for four certification types (Better Cotton Initiative (BCI), Cotton made in Africa (CmiA), Fairtrade Labelling Organization (FLO), and EU Organic Regulation) [21]. Targosz-Wrona provides an overview of the label requirements on chemical residues in fibres and emissions thresholds for the textile manufacturing phase for the labels EU Flower, Ecological product, Eco-sign, Slovak environmental friendly product, and Nordic Swan [22]. Henniger compares the requirements of the 15 most relevant textile labels for the UK market with regard to different environmental aspects (e.g., water use, deforestation, CO_2-emissions) and assessment approaches adopted by the labels (e.g., life cycle assessment, raw material assessment) [23]. An analysis of the labels requirements considering all life cycle stages of textile products is carried out by Clancy et al. for six ecolabels with high relevance for the market in Sweden (EU Ecolabel, Bluesign, Cradle-to-cradle, Made-by, Textile Exchange, Oeko-Tex) [24]. For each life cycle stage from design and raw material production to waste management, the authors evaluate whether the labels provide specific requirements (e.g., a restriction of the use of specific chemicals) or optional/indirect criteria (i.e., the requirement is not binding or the life cycle stage is influenced by the requirements for a different life cycle stage). An analysis of the complete life cycle of textiles, as well as different environmental aspects and hotspots, is conducted in the study of Minkov et al., who compare similarities and gaps between the requirements of the Product Environmental Footprint (PEF) and European Flower (EUF) [25]. Although all aforementioned studies evaluate label requirements, two following questions remain unclear: (1) Whether the focus of the labels with regard to considered environmental aspects and life cycle stages of products is comparable and (2) whether the label requirements address the main environmental hotspots in the life cycle of textiles, and thus contribute to the reduction of the environmental burden of certified products.

To address this gap, the goal of this paper is to evaluate similarities and gaps between textile ecolabels and analyse their focus areas concerning covered environmental aspects and life cycle stages. For this, seven textile ecolabels with different scopes and approaches for the requirement setting are evaluated with regard to their main characteristics (e.g., type of communication, scope, etc.), addressed environmental issues (e.g., climate change, water use, etc.), and covered life cycle stages of textiles. Following labels were selected for the analysis: Blue Angel Textiles [26], bluesign® [27], Cotton made in Africa (CMiA) [28], Cradle to Cradle Certified™ [29], Global Organic Textile Standard (GOTS) [13], Global Recycled Standard (GRS) [30], and VAUDE Green Shape [31]. Further, we analyse whether the environmental requirements of the labels cover the hotspots with regard to environmental aspects and life cycle stages of textiles. This paper addresses only environmental aspects of sustainability, omitting other sustainability dimensions (social and economic criteria) of ecolabels. It is structured as follows: Section 2 provides an overview of the life cycle stages and environmental hotspots of textiles and introduces the characterization scheme for the ecolabels, in Section 3, a description of the selected ecolabels and methodological procedure is provided, Section 4 presents the results. In Section 5, the results are discussed, and Section 6 concludes with a short outlook.

2. Theoretical Background

2.1. Environmental Impacts throughout the Textile Life Cycle

Based on existing literature (see Table S1), five main life cycle stages of textiles can be identified:

- Raw material production;
- textile manufacturing;
- distribution;
- garment use;
- textile disposal.

The raw material production phase considers either the growing of natural fibres such as cotton, wool, silk, and flax, or the manufacturing of fibres made from a variety of raw material sources, including plant, animal, and synthetic polymers [32]. The main concerns in this stage originate from either the agricultural production and the attributed intense use of water and pesticides or the production of synthetic and cellulosic fibres and the resulting emissions to air and water [32]. One of the most famous examples of the severe environmental consequences that can occur through cotton cultivation is the tragedy of the Aral Sea. The increased water diversion for irrigation of cotton fields lead to an insufficient water supply from its two river sources, causing the Aral Sea to dramatically decrease in size and water volume since the early 1960s [33].

The yarn and textile manufacturing itself has several steps including sizing, knitting, pre-treatment, dyeing, and finishing. The making up process encompasses, pattern drafting, producing samples, cutting, sewing, and applying embellishments [34,35]. The environmental issues in this phase vary from the inhalation of cotton dust during the yarn manufacturing, to the contamination of wastewater with mineral knitting oils, remaining pesticides, and leftovers from bleaching, as well as dyes that usually contain heavy metals and auxiliary chemicals used for finishing. For the distribution phase, the garments are usually packed in polyester bags and distributed to warehouses or retailers [35].

The garment use phase is characterized by acquisition, use, and maintenance activities [34]. It is mainly concerned with washing and drying the garments. Thus, the environmental impacts are associated mainly with electricity, detergent, and water use [36]. The nature and quality of a fibre can further influence the maintenance of a textile [37]. The quality of cotton fibres, where high quality fibres are not as easy to get dirty, as well as the difference between mechanical and chemical treatment, can significantly impact the behaviour of the fabric in use [37].

During the textile disposal phase, sending the apparel to landfills dominates re-use, recycling, and other end-of-life management activities [34].

2.2. Textile Ecolabels

During the past decades, increasing attention of the consumers to the environmental and social impacts of products resulted in an increasing adoption of sustainability practices in business, e.g., eco-innovation and lean management. The latter allow companies to reduce the environmental burden associated with their production activities, and at the same time to foster the development of new products, technologies, or business structures, which increases their overall market viability [38]. As demonstrated in recent studies, the implementation of Corporate Social Responsibility (CSR) strategies gained importance for the competitiveness in the textile sector [39–42]. One of the strategic CSR areas is the so-called "marketplace CSR", which includes company's communication with its suppliers, consumers, and other stakeholders along the value chain. Particularly with regard to the consumer relations, textile producers increasingly adopt ecolabels to demonstrate (improved) environmental and/or social performance of their goods [41].

Ecolabels are voluntary environmental product information schemes (EPIS), which are used in order to systematically approach the environmental information of a product.

The mandatory approach to EPIS includes declarations of contents such as food ingredients, usage, and disposal information, mainly applying to chemical substances and products. The voluntary approach to EPIS (i.e., ecolabels) leaves it to the market actors to decide whether to sign or label their product. In the following, the focus is set on the voluntary ecolabels and declarations. The overall goal of the voluntary environmental labels and declarations is encouragement of the demand for and supply of the products that cause less pressure on the environment. This is achieved through communication of verifiable and accurate information on the product's environmental performance [43]. Stø et al. [44] demonstrated that product information is usually asymmetrically allocated between buyers and sellers. This knowledge gap can only be filled through external support as supposedly offered by ecolabels and EPIS [44].

The ecolabel or environmental declaration should consider the life cycle of a product or service from production to final disposal. However, the undertaking of a life cycle assessment is not always necessarily required [43]. Three types of environmental labelling are further specified by the ISO standards: Environmental labels (Type I), self-declared environmental claims (Type II), and environmental declarations (Type III) [45–47].

The first voluntary public ecolabels were developed following the introduction of the German Blue Angel label in the 1970s [11,48], which provided information about products with the best environmental characteristics in the entire life cycle of a product [11]. They were followed in the next years by a proliferation of eco-labelling and single-issue certification, as well as the development of individual company private standards [11,48]. Since the 2000s, a large number of ecolabels and other ELIS coexist [11,48].

2.3. Characterization Scheme for Environmental Labels and Declarations

As described in the previous section, three types of environmental labels and declarations are distinguished according to ISO. Nevertheless, as demonstrated in recent studies, several ecolabels cannot be assigned to any of these types due to different awarding criteria and formats, which makes it difficult to classify and compare ecolabels [49]. A recently introduced characterization scheme overcomes this obstacle by introducing 22 attributes with regard to following aspects of the labels: communication, scope, standard characteristics, governance, and conclusive characteristics [18,49]. In the following, these attributes and some examples of corresponding label features are shortly introduced. A detailed description of all characterization attributes and features can be found in the study of Minkov et al. [18,49].

The aspect communication characteristics includes the following five attributes: ISO typology (e.g., Type I, undefined), awarding format (seal, rating), multiplicity of covered aspects (single or multi-aspect), aspects diversity (environmental, social), and end-user focus (e.g., business-to-business (B2B)). The aspect scope includes the attributes sector scope (i.e., sector-specific or multi-sector), operational scope (e.g., product, organization), geographical scope (national, international), awarding criteria scope (product-specific or generic), application of materiality principle, and life cycle perspective. The aspect standard characteristics considers compulsoriness (voluntary or mandatory), financing, purpose (i.e., idealistic or neutral), and longevity (single issued or renewable). The aspect governance characteristics includes the attributes governance (governmental, private), verification (e.g., first or second party), awarding criteria revision, and stakeholder involvement (low, high). The aspect conclusive characteristics consists of three attributes: Transparency, comparability, and environmental excellence.

3. Materials and Methods

3.1. Selected Ecolabels

As stated in the introduction of the paper, this study aims at analysing textile ecolabels with different scopes and approaches for setting the requirements on the environmental issues. In the following, the reasons for the inclusion of each label in this study are explained, and the labels are shortly introduced. Table 1 summarizes the general information of the ecolabels.

The seven ecolabels were selected considering their relevance as an ecolabel as well as their relevance for their individual focus area (i.e., cotton production, circularity, recycling). The Blue Angel Textile label was chosen due to the label's relevance as the oldest existing ecolabel. The bluesign ecolabel has a strong focus on chemical use and is considered to be one of the strictest ecolabels in this area. The Cotton made in Africa ecolabel has a regional validity for sub-Saharan Africa and is one of the most relevant organic labels with a focus on cotton with many corporate labels referring to it. The Cradle to Cradle Certified™ ecolabel set a clear focus on circularity and is relevant, as the ecolabel requirements are specifically based in the Cradle to Cradle concept. The Global Organic Cotton Standard ecolabel proves its relevance as one of the most commonly used and best known ecolabels. The Global Recycling Standard is relevant within the special focus area of recycling. The VAUDE Green Shape ecolabel was chosen for this analysis as a company initiated ecolabel that was possible to analyse due to its comparably well provided information on the ecolabels criteria.

3.1.1. Blue Angel Textiles

The Blue Angel Textiles label was established in a cooperation of the Federal Ministry for the Environment, Nature Conservation, and Nuclear Safety and the German Environmental Agency. The objective of the label is to offer guidance for sustainable products through four approaches: "Promoting higher environmental standards in the production process; improving occupational safety and social conditions during production; avoiding chemical hazards to health in the end product; verifying the product's fitness for use" [50].

3.1.2. Bluesign®

Under the name bluesign® system, bluesign technologies AG created a network of chemical suppliers, manufacturers, and brands which are guided by the bluesign® criteria. The bluesign® system covers all bluesign® criteria, and the bluesign® system partners based on the management of inputs and responsible actions across the whole supply chain following five principles: Resource productivity, consumer safety, water emissions, air emissions, and occupational health and safety [27]. When being awarded the bluesign® label, all involved parties need to follow certain milestones, for example, a bluesign® system partner agreement, the certification of chemical products and articles, as well as labelling [51]. The end-product is labelled a bluesign® product if at least 90% of the used fabric and at least 30% of the used accessories are bluesign® approved [52]. Part of the bluesign® system is the bluesign® system substances list. It includes around 900 substances that are either not permitted (around 600) or subject to certain limitations. Within the bluesign® system, chemicals are rated as blue, grey, or black. Blue rated chemicals fulfil all criteria for the final product, the worker, and the environmental release. Grey rated chemicals can only be used under certain conditions for bluesign® approved materials, while black rated chemicals fail the criteria and their use is not accepted.

Table 1. General information on selected labels.

Name	Focus	Short Description	Managing Organisation	Founded in	Reason for Selection
Blue Angel Textiles	Textile products	Being the oldest environmental label the Blue Angel aspires to provide reliable guidance for consumers. The Blue Angel Textiles represents a subcategory of the Blue Angel Label which certifies a wide range of products.	Federal Ministry for the Environment, Nature Conservation and Nuclear Safety; German Environmental Agency; Environmental Label Jury; RAL gGmbH	1978	Oldest existing ecolabel
bluesign®	Textile manufacturing chain/ecological footprint	With special focus on the used chemicals, bluesign® offers a standard for suppliers, manufacturers, and top-brands to reduce their textile footprint.	Bluesign Technologies AG	2000	Considered to be one of the strictest ecolabels regarding chemical use
Cotton made in Africa (CmiA)	Cotton	The label promotes sustainable cotton growing and farming approaches to enable African cotton farmers to improve their living conditions on their own, referring to ecological, social, and economic aspects.	Aid by Trade Foundation (ABTF) Non-profit	2005	One of the most relevant organic label with a focus on cotton; regional focus for sub-Saharan Africa
Cradle to Cradle Certified™	Circularity	Based on the Cradle to Cradle framework, the certificate consists of basic, silver, gold, and platinum levels for safer, more sustainable products made for the circular economy. The label certifies different product categories one of which is textiles.	Cradle to Cradle Products Innovation Institute Non-profit	2005 by McDonough Braungart Design Chemistry (MBDC) and donated to the Cradle to Cradle Products Innovation Institute in 2010.	One of the most relevant labels with a focus on circularity
Global Organic Textile Standard (GOTS)	Organic natural Fibre Products	The self-declared leading textile standard considers social and environmental criteria in the processing of organic fibres throughout the entire textile supply chain.	Global Standard gGmbH	2005 agreement on the first version and implementation scheme.	One of the most commonly used and best known textile ecolabels
Global Recycled Standard (GRS)	Recycling	Observing the full supply chain, the standard focuses on traceability, environmental principles, social requirements, and labelling. It tracks and verifies recycled input material from input to the final product.	Textile Exchange Non-profit	2008	One of the most relevant labels with a focus on recycling
VAUDE Green Shape	Functional, environmentally friendly textile products	The corporate label certifies its own textile products made from sustainable materials covering the whole life cycle of the product.	VAUDE Sport GmbH & Co. KG	2009	A company initiated label that provides information about the certification requirements

3.1.3. Cotton made in Africa (CmiA)

The Cotton made in Africa label was designed by the Aid by Trade Foundation with the goal to improve living conditions for local farmers and promote environmentally friendly cotton production [28]. The criteria set for the CmiA is two-tier. The first set includes criteria that determine if farmers and companies can participate in the program. The second-level criteria are sustainability criteria. The participants in the CmiA programme are not immediately required to meet all sustainability criteria, but can develop and improve following a development plan. The criteria follow a traffic light assessment that rates the status of the criteria as green, yellow, or red [53]. For the entry phase, a minimum of 50% of the sustainability criteria must be rated as yellow or green. All red and yellow classified sustainability criteria must have recommendations for possible improvement. In the next verification after two years, in an ideal case all formerly red criteria are improved to yellow and the yellows to green. For subsequent verifications, ideally all criteria should now be rated green and the overall green status should be maintained [28].

3.1.4. Cradle to Cradle Certified™

The Cradle to Cradle approach integrates multiple attributes, such as safe materials, continuous reclamation and reuse of materials, clean water, renewable energy, and social fairness [29]. A decisive aspect in the Cradle to Cradle approach is the definition of the three principles: Eliminating the concept of waste, use renewable energy, and celebrate diversity [54]. The goal is to achieve a perpetual cycling of ingredients which either biodegrade naturally and restore the soil or are being fully recycled into high quality materials for subsequent product generations. Cradle to Cradle therefore defines two effective material cycles: The biological cycle, able to safely re-enter the biological system, and the technical nutrient cycle, where products or materials can be recovered at the end-of-use phase [54]. This approach has been criticized by many scholars due to its theoretical nature and lacking feasibility [19]. The Cradle to Cradle™ label applies to materials, sub-assemblies, and finished products. To create a standard that promotes improvement, the label uses a 5-Level System of Basic, Bronze, Silver, Gold, and Platinum. In order to qualify for one of the levels, the requirements from all lower levels must be met as well. The final certification level is determined by the minimum level of achievement in the five different levels [54].

3.1.5. Global Organic Textile Standard (GOTS)

The GOTS standards was initiated in 2002 at the Intercot Conference and was started as a certification system in 2006. Its aim is to ensure an organic status of textiles from harvesting through socially and environmentally responsible manufacturing up to labelling. In recognition of the fact that textile production today is nearly impossible without chemicals, the label defines criteria for low impact and low residual natural and synthetic chemical inputs [55]. The standard offers two label grades either "organic"/"organic—in conversion" or "made with (x%) organic materials"/"made with (x%) organic materials—in conversion" [55]. The criteria focus on compulsory criteria with only expressly stated exceptions.

3.1.6. Global Recycled Standard (GRS)

The Global Recycling Standard, initiated by Control Union, was passed on to Textile Exchange in 2011, who also own and administer other standards such as the Content Claim Standard (CCS) and the Recycled Claim Standard (RCS). The overall goal of the GRS is to increase the use of recycled materials in products while reducing or eliminating the harm caused by their production. It aims to concentrate on recycled content, the chain of custody, social and environmental practices, as well as chemical restrictions [56]. The GRS can be used for any product that contains at least 20% recycled materials [56].

3.1.7. VAUDE Green Shape

The Green Shape Label is the corporate label from the outdoor outfitter VAUDE. It was invented by the company due to the absence of a comprehensive textile label [57]. With the Green Shape Label, VAUDE claims to have "developed its own rating system for environmentally friendly outdoor products" [57]. According to VAUDE's online presentation of the label, it "covers the entire product lifecycle with its strict standards—from design and production to maintenance, repair, and disposal" [57].

3.2. Analysis of the Labels

First, a characterization of the selected labels is carried out based on the characterization scheme proposed by Minkov et al. (see Section 2.3) [18,49]. Next, we analyse the label requirements following a three-step procedure. In the first step, considered environmental aspects (e.g., water use) and life cycle phases (e.g., raw material production) were identified based on the documentation of the labels. Then, the label requirements were assigned to the life cycle stages and environmental aspects of textile products. If a requirement could not be assigned to one specific environmental aspect (e.g., the prerequisite to use organic materials influences several environmental aspects including toxicity, water use, and land use), it was identified as a "general" requirement (see Table 2).

Table 2. Exemplary table for the analysis of ecolabels.

Life Cycle Step/Env. Aspect	Toxicity	Water Use	Air Emissions	Land Use	Recycling
Raw material production	General requirements.				
	Specific requirements.	Specific requirements.	Specific requirements.	Specific requirements.	Specific requirements.
Textile Manufacturing					
Distribution					
Garment Use					
Textile Disposal					

Finally, we compare the requirements of the labels to the environmental hotspots that occur in the life cycle of textiles following the procedure proposed by Minkov et al. [25]. This is done based on the hotspots analysis published as part of the Product Environmental Footprint Category Rules (PEFCR) for t-shirts [58]. The latter were developed within the Product Environmental Footprint (PEF), which aims at providing a harmonized methodology and rules for the environmental assessment of products under the life cycle perspective [59,60]. The PEF study provides an overview of the environmental hotspots on a level of impact categories (e.g., climate change), life cycle stages (e.g., production of material), and processes (e.g., cotton fibres) with the cradle-to-grave system boundary. The results of the PEF study [58] were considered for the impact categories that relate to the environmental impacts with a high relevance in the life cycle of textiles (see Section 2.1): Climate change (impact on air emissions), water scarcity (impact on water consumption), acidification (terrestrial and freshwater), and freshwater eutrophication (impact on water pollution) (see Table 3).

The applied methodological procedure is illustrated in Figure 1.

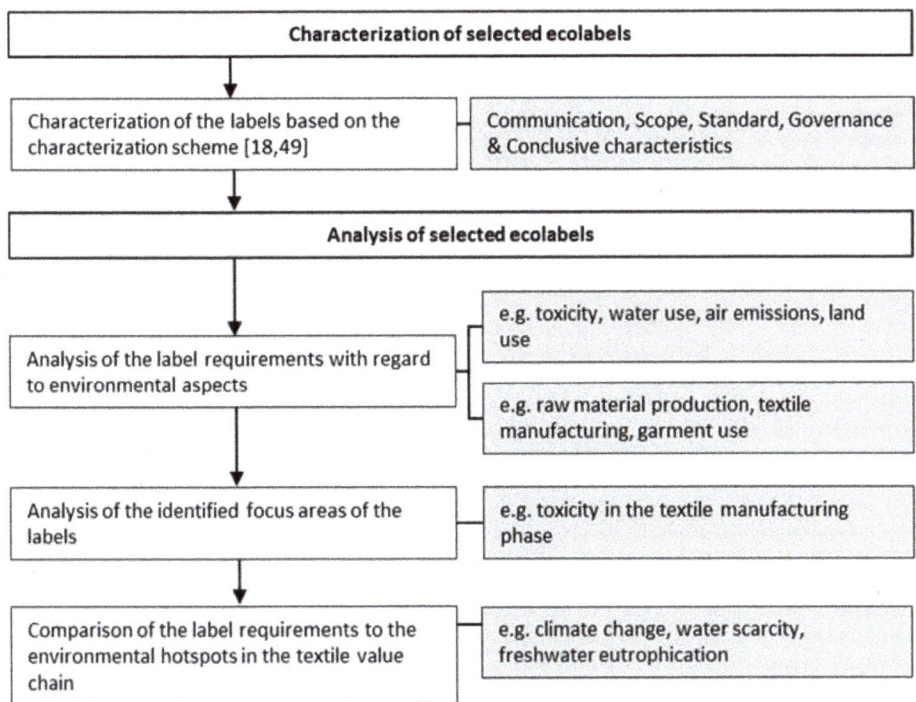

Figure 1. Methodological procedure. White boxes indicate working steps carried out within the methodological procedure; grey boxes demonstrate examples of the outcomes of each step.

Table 3. Overview of environmental hotspots of a T-shirt for selected impact categories over the life cycle (modified from [58]).

		% Contribution			
Life Cycle Stage	Processes	Climate Change	Water Scarcity	Acidification (Terrestrial and Freshwater)	Freshwater Eutrophication
Production of material	Cotton fibres	23.7	58.3	24.6	75.7
Production of T-shirt	Spinning, production of cotton yarn (combed)	5.0	-	6.8	-
	Spinning, production of cotton yarn (carded)	3.3	-	4.7	-
	Circular knitting	3.6	-	5.1	-
	Fabric dyeing	10.7	-	11.3	-
	Yarn dyeing	3.9	-	3.9	5.0
	T-shirt assembly	9.1	-	14.2	-
	Total production	35.6	-	46.0	5.0
Transportation by customer	Passenger car, average	13.6	-	7.2	-
Use stage/Washing	Electricity grid mix 1 kV–60 kV	7.9	30.0	4.0	-

4. Results
4.1. Characterization of Selected Ecolabels

The results of the applied characterization scheme are shown in Table 4. All analysed labels show several similarities regarding the communication characteristics, more specific all have a multi-aspect approach, address both environmental and social and/or health aspects, and have a B2C focus. Five labels represent a seal, while CmiA and Cradle to Cradle^{TM} label follow a rating awarding format. A significant difference between the labels can be detected for the attribute ISO typology. Only the Blue Angel Textiles and GOTS label are a fully conformant Type I eco-label program. The rest of the labels does not fully conform with the Type I requirements, and the typology of the CmiA label can be characterized as "undefined". With regard to the sectoral scope, three labels can be characterized as multi-sectoral (Blue Angel Textiles as part of the Blue Angel label, Cradle to Cradle^{TM}, and GRS), while all other labels serve for the textile products only (or cotton in case of the CmiA label). Except for the CmiA, which is applicable only for the cotton production in Africa, all labels have an international geographical scope and claim to apply the life cycle perspective by providing requirements for different life cycle stages of textiles, e.g., raw material production, textile manufacturing, and use. This attribute is analysed in detail in the next chapters.

The labels show similarities also with regard to the attribute standard characteristics, for example, all labels are voluntary and ideals-centric, i.e., serve as a benchmark of achieving certain ideals or excellence. In contrast to the VAUDE Green Shape, which is a single-issued label (i.e., is never re-verified), all other labels are renewable (are revised and reissued after expiration) or improvement-based (CmiA and Cradle to Cradle^{TM}), which means that they require a demonstration of improved performance for a re-certification [18].

The Blue Angel Textiles is the only one quasi-governmental label (i.e., initiated by a government, but managed by a private company), while other labels are private. Other governance characteristics are addressed similarly by all labels except VAUDE Green Shape, e.g., the labels are verified by third party, have regularly revised awarding criteria and medium to high stakeholders involvement. The VAUDE Green Shape, in contrast, is second party certified (verification through VAUDE Sports) and does not provide information on the attributes awarding criteria revision and stakeholders involvement.

All analysed labels have a high level of transparency (only for the VAUDE Green Shape, the program rules cannot be accessed) and intend environmental excellence (i.e., the certification promotes environmental excellence of the product). Five labels have a medium score for the characterization attribute comparability, since these labels do not allow a comparison between products awarded by the same scheme, but intend superiority to non-awarded products. The comparability of the CmiA and Cradle to Cradle^{TM} labels is evaluated as low, since the comparison of products is difficult due to different levels of conformity introduced by these labels.

Table 4. Characterization of the labels according to the characterization scheme by Minkov et al. (2018, 2019).

Attribute	Blue Angel Textiles	Bluesign®	Cotton Made in Africa (CmiA)	Cradle to Cradle Certified™	Global Organic Textile Standard (GOTS)	Global Recycled Standard (GRS)	VAUDE Green Shape
Communication Characteristics							
ISO typology	Fully conformant Type I eco-label program according to ISO 14024	Does not fully conform with Type I requirements of ISO	Undefined	Does not fully conform with Type I or Type III label requirements of ISO	Fully conformant Type I eco-label program according to ISO 14024	Does not fully conform with Type I requirements of ISO	Does not fully conform with Type I requirements of ISO
Awarding format	Seal	Seal	Rating (sealed; ranked on a traffic light system)	Rating (sealed; ranked on a predefined scale after complying with minimum performance criteria)	Seal	Seal	Seal
Multiplicity of covered aspects	Multi-aspect	Multi-aspect	Multi-aspect	Multi-aspect	Multi-aspect	Multi-aspect	Multi-aspect
Aspects diversity	Mostly environmental and occupational health and safety	Mostly environmental and occupational health and safety	Both environmental and social/health	Both environmental and social/health	Both environmental and social/health	Both environmental and social/health	Both environmental and social/health
End-user focus	B2C	B2C	B2C	B2B and B2C	B2C	B2B und B2C	B2C
Scope							
Sector scope	Multi-sectorial	Sector specific	Sector specific	Multi-sectorial	Sector-Specific	Multi-Sectoral	Sector-Specific
Operation scope	Product	Product	Product	Product (certain criteria in three of five quality categories relate to the organization)	Product	Product	Product
Geographic scope	International	International	Regional (Africa)	International	International	International	International
Awarding criteria scope	Product specific (specific requirements for different fibres)	Generic	Generic	Generic (equal criteria for all products)	Generic	Generic	Product specific (different requirements for apparel and other textile products)
Materiality principle	No (all products are assessed against the same set of criteria, independent from their individual materiality)	No (all products are assessed against the same set of criteria, independent from their individual materiality)	No (all products are assessed against the same set of criteria, independent from their individual materiality)	No (all products are assessed against the same set of criteria, independent from their individual materiality)	No (all products are assessed against the same set of criteria, independent from their individual materiality)	No (all products are assessed against the same set of criteria, independent from their individual materiality)	No (all products are assessed against the same set of criteria, independent from their individual materiality)

Table 4. Cont.

Attribute	Blue Angel Textiles	Bluesign®	Cotton Made in Africa (CmiA)	Cradle to Cradle Certified™	Global Organic Textile Standard (GOTS)	Global Recycled Standard (GRS)	VAUDE Green Shape
Standard Characteristics							
Life cycle (LC) perspective	LC based	Partly LC based	Non-LC based	Partly LC based	LC based	Partly LC based	LC-based
Compulsoriness	Voluntary	Voluntary	Voluntary	Voluntary	Voluntary	Voluntary	Voluntary
Financing	Governmental subsidies	Information not provided	Fees and/or member dues; donations	Fees and/or member dues	Self-financed using yearly licence fees and certification costs	Fees and/or member dues; other (Consulting, etc.)	Information not provided
Purpose	Ideals-centric	Ideals-centric	Ideals-centric	Ideals-centric (a benchmark of achieving conformance with the C2C principles)	Ideals-centric	Ideals-centric	Ideals-centric
Longevity	Renewable Label validity: Three to five years	Renewable Label	Improvement-based; valid for two years	Improvement-based (in case of re-certification, intentions for improvement must be reported)	Renewable	Renewable	Single-issued
Governance Characteristics							
Governance	Quasi-governmental	Private	Private	Private	Private	Private	Private
Verification	Third party (mandatory by independent, external body)	Third party	Third party	Third party (mandatory by independent, internal certification body; however, independence of the conformance assessment body not assured)	Third party	Third Party	Second party (verification through VAUDE Sports)
Awarding criteria revision	Yes, regularly; criteria revised after three to five years	Yes, regularly; criteria revised at least every four years	Yes regularly (interval of revision cycle not clear)	Yes, regularly (revision of the Product Standard to be done every three years)	Yes regularly; criteria revised after three years	Yes regularly (interval of revision cycle not clear)	Information not provided
Stakeholders involvement	High (open consultations during the development of new or updating existing awarding criteria)	High (public consultation during the revision of bluesign® criteria)	High (open consultation during revision of the standards)	Medium (during the product standard revision process, two public comment periods are at disposal for comments by stakeholders; not yet carried out in practice)	High (selected stakeholders invited to participate in the revision process)	High (submission of feedback always possible)	Information not provided

Table 4. Cont.

Attribute	Blue Angel Textiles	Bluesign®	Cotton Made in Africa (CmiA)	Cradle to Cradle Certified™	Global Organic Textile Standard (GOTS)	Global Recycled Standard (GRS)	VAUDE Green Shape
			Conclusive Characteristics				
Transparency	Program rules—yes Awarding criteria—yes Awardees—yes	Program rules—yes Awarding criteria—yes Awardees—yes	Program rules—yes Awarding criteria—yes Awardees—yes	Program rules—yes Certification criteria—yes Awardees—yes	Program rules—yes Certification criteria—yes Awardees—Yes	Program rules—yes Certification criteria—yes Awardees—yes	Program rules—no Certification criteria—yes Awardees—Yes
Comparability	Medium (comparison and comparative assertions are not possible between products awarded the same label; awarded products can claim superiority to non-awarded products)	Medium (binary awarding system prohibits comparability between products awarded the ecolabel; awarded products can claim superiority to non-awarded products)	Low (comparison between products is difficult due to the different levels of conformity based on the traffic light system colours)	Low (comparison between products is difficult due to the five quality categories; comparative assertions are not possible; comparability is not strived for by the program)	Medium (binary awarding system prohibits comparability between products awarded the ecolabel; awarded products can claim superiority to non-awarded products)	Medium (binary awarding system prohibits comparability between products awarded the ecolabel; awarded products can claim superiority to non-awarded products)	Medium (binary awarding system prohibits comparability between products awarded the ecolabel; awarded products can claim superiority to non-awarded products)
Environmental excellence	Intended	Intended	Intended	Intended (however, frontrunner principle not applied)	Intended	Intended	Intended

4.2. Considered Environmental Aspects and Life Cycle Phases

In the following, the results with regard to the considered environmental aspects and life cycle phases are presented (see Table 5 and Tables S2–S8).

The Blue Angel Textiles label provides requirements for all life cycle stages from raw material production to distribution, while the use phase and disposal are not considered. The raw material production stage is considered most extensively compared to other labels, since all impact categories are addressed and also general requirements are provided. The textile manufacturing stage is also considered by means of both general and specific requirements. While a comprehensive requirements set is provided for the toxicity, water use, and air emissions, two other aspects (land use and recycling) are not addressed in this stage. For the distribution, few requirements with regard to toxicity, recycling, and land use are provided.

The bluesign® label addresses two life cycle stages of textiles: Raw material production and textile manufacturing. For the raw material production, a set of general requirements is provided, e.g., that all raw materials used must be bluesign® approved. For the textile manufacturing stage, both general requirements (e.g., availability of a management system with a plan-do-check-act cycle covering quality, environment/resource savings, and occupational health and safety) and specific requirements for all environmental aspects are provided. Quantitative thresholds are given for the impacts on toxicity, water use, and air emissions, while for land use and recycling, qualitative targets are provided (e.g., "packaging shall be reduced ... ").

The CmiA label is designed for only the cotton production stage, therefore it provides requirements only for the raw material production, while other life cycle stages of textiles are not considered. The label provides both general and specific requirements, while the level of conformity can be achieved on three levels: Red (non-conformity), yellow (partly conformity), and green (full conformity). Furthermore, excluding criteria are provided, e.g., use of pesticides banned under the Stockholm Convention on Persistent Organic Pollutants (POPs), cotton production under irrigation and cutting of primary forest. The requirements set by the label are mainly quantitative, e.g., sufficient evidence of the risks and dangers related to the storage of pesticides and application of methods for water conservation.

The Cradle to Cradle Certified™ label has a strong focus on the textile manufacturing step. In this step toxicity, water use, air emissions, as well as recycling are addressed, while only the impacts on land use are not considered. The label requirements follow a 5-Level System, which sets basic, bronze, silver, gold, and platinum criteria. The differentiation between basic and platinum criteria is vast and distinct: While for water use in the textile manufacturing, the basic criteria requires no significant violation of discharge permit within the last two years, the platinum criteria requires that only water that meets drinking water quality may leave the manufacturing facility. While raw material production, distribution, and the garment use phases are not addressed at all, in the textile disposal phase, requirements address the environmental aspect recycling.

The GOTS label addresses the raw material production phase with general criteria, i.e., requirement on the share of the fibres produced as "organic". The textile production phase addresses toxicity, water use, and air emission. For the environmental aspects land use and recycling, no requirements were identified in this life cycle step. In contrast to the Cradle to Cradle Certified™ label, GOTS does not set different certification levels. The requirements are presented as general requirements as well as in relation to the individual production steps such as dying, printing, and finishing or sizing and wet processing stages. In the distribution phase, environmental aspects toxicity, air emissions, and land use are covered, while the garment use phase and the textile disposal phase are not considered.

Table 5. Overview of considered life cycle steps and environmental aspects. The colours indicate hotspots in the life cycle stages and environmental impacts according to the PEFCR: Orange—over 20% of the total impact, yellow—over 10% of the total impact. It should be noted that the PEFCR hotspot data was not available for the impacts toxicity, land use, and recycling.

Life Cycle Step/env. Aspect		Toxicity	Water Use	Air Emissions	Land Use	Recycling
Raw material production	General requirements			Blue Angel Textiles bluesign® Cotton made in Africa (CmiA) Global Organic Textiles Standard (GOTS) VAUDE Green Shape		
	Specific requirements	Blue Angel Textiles Cotton made in Africa (CmiA) Global Recycling Standard (GRS)	Blue Angel Textiles Cotton made in Africa (CmiA)	Blue Angel Textiles	Blue Angel Textiles Cotton made in Africa (CmiA)	Blue Angel Textiles Global Recycling Standard (GRS)
Textile Manufacturing	General requirements			Blue Angel Textiles bluesign® Global Recycling Standard (GRS) VAUDE Green Shape		
	Specific requirements	Blue Angel Textiles bluesign® Cradle to Cradle Certified™ Global Organic Textiles Standard (GOTS) Global Recycling Standard (GRS) VAUDE Green Shape	Blue Angel Textiles bluesign® Cradle to Cradle Certified™ Global Organic Textiles Standard (GOTS) Global Recycling Standard (GRS)	Blue Angel Textiles bluesign® Cradle to Cradle Certified™ Global Organic Textiles Standard (GOTS) Global Recycling Standard (GRS)	bluesign®	bluesign® Cradle to Cradle Certified™ Global Recycling Standard (GRS)
Distribution	Specific requirements	Blue Angel Textiles Global Organic Textiles Standard (GOTS)	Not addressed	Global Organic Textiles Standard (GOTS)	Blue Angel Textiles Global Organic Textiles Standard (GOTS)	Blue Angel Textiles
Garment Use	Specific requirements	VAUDE Green Shape	Not addressed	VAUDE Green Shape	Not addressed	Not addressed
Textile Disposal	Specific requirements	Not addressed	Not addressed	Not addressed	Not addressed	Cradle to Cradle Certified™

The GRS label addresses the raw material production and textile manufacturing phases. In the raw material production phase, only the environmental aspects of toxicity and recycling are considered. For the textile manufacturing phase, both general requirements (e.g., Certified Organizations are required to have an environmental management system) and specific requirements (e.g., water use: A drainage plan with understanding of wastewater flow direction and discharge point is required) are provided. The latter consider all environmental impacts except land use and are mainly quantitative, e.g., the rules on the use and storage for chemicals and monitoring of emissions.

In contrast to other labels analysed in this research work, the VAUDE Green Shape label considers besides the raw material production and textile manufacturing the use phase of the garment. For the raw material production, only one criteria is provided, which prohibits any usage of GMO. The general requirements for textile manufacturing include prohibition and rules for the usage of some chemicals (e.g., motif prints need to be either water based or based on sublimation) and the requirement that a minimum of 90% of used garment must be certified/declared. A broad range of certification options is provided, which include supplier certification (e.g., ISO 14001, EMAS), fabric certification (e.g., bluesign® approved, GOTS), or "eco-fabric" (e.g., organic cotton, TENCEL, chlorine free wool). Furthermore, specific requirements for toxicity are provided, according to which compliance with the manufacturing restricted substance list (MRSL) must be assured. In the textile use phase, environmental impacts on toxicity (high impact care) and air emissions (the product requires tumble drying, i.e., high energy use and impact on climate change) are addressed.

Overall, it can be summarized that the Blue Angel Textiles label covers most life cycle phases in the considered environmental impact categories. Followed by a wide margin, the GRS and GOTS label also take into account several life cycle phases.

4.3. Overview of Identified Focus Areas of Selected Labels

In the following section, the identified requirements for the environmental impacts and life cycle phases are presented (see Table 5).

Looking at the life cycle steps, most requirements are formulated for the life cycle stages raw material production and textile manufacturing. For each life cycle step, both general criteria and criteria specific to environmental aspects exist. In the raw material production step, most labels set only general criteria, i.e., requirements on general cultivation practices, for example, controlled organic cultivation (Blue Angel Textiles, GOTS) or chemicals, particularly pesticides management (GOTS, bluesign®). The Blue Angel Textile label addresses all specific environmental aspects, e.g., by providing thresholds for the content of specific pollutants present in the fibres. CmiA addresses specific environmental aspects including toxicity, water use, and land use. The GRS label addresses toxicity (restriction of certain chemicals) and recycling (i.e., recycling content).

The textile manufacturing is extensively addressed by all evaluated labels. Most labels provide general criteria, which include requirements on environmental management systems (GRS, bluesign®) or overall compliance of all manufacturing processes with the local legislation at the production site (GRS, Blue Angel Textiles). Specific criteria, for example, thresholds for application of chemicals and wastewater quality parameters are also provided by most labels.

Significantly less focus is set on the distribution, garment use, and textile disposal phases. Only four of the seven ecolabels address these steps, and no ecolabel set any general criteria. In the textile disposal phase, none of the environmental aspects are addressed apart from recycling by the Cradle to Cradle Certified™ label.

Regarding the addressed environmental aspects, toxicity has a clear dominance, and is covered by all seven ecolabels. The aspects of water use and air emissions are addressed by six ecolabels, while the aspects of land use and recycling are addressed by only four ecolabels. The differences in focus on the environmental aspects are not as extreme as the differences in the life cycle steps.

4.4. Comparison of the Label Requirements and Environmental Hotspots Identified by PEF

The identified label requirements are compared with the environmental hotspots identified by PEF. The comparison was performed based on the PEF study for the impacts water use and air emissions (see Section 3). For other impacts addressed by the labels and analysed in this work (toxicity, land use, and recycling), no hotspot data was available. The hotspot in the impact water use was identified based on the impact categories water scarcity (i.e., water consumption) as well as acidification and freshwater eutrophication (i.e., water pollution). Only two labels—Blue Angel Textiles and CmiA—provide specific requirements for water use in the raw material production stage, whereas only CmiA considers water consumption, e.g., by prohibiting cotton production under irrigation and setting goals for the application of water conservation techniques. A clear environmental hotspot with regard to water pollution occurs in the textile manufacturing phase. Here, all analysed labels (except CmiA that considers only raw material production phase) provide requirements with regard to the quality of discharged water, e.g., by setting thresholds for specific pollutants or requiring compliance with local legislation. In contrast to material production and textile manufacturing phase, water use aspects in the garment use phase are not addressed by any of the analysed labels, although this stage contributes to one-third of the total water scarcity impact in the life cycle of textiles (see Table 3). The hotspot for the impact on air emissions was identified based on the impact category climate change considered in the PEFCR. Still, it should be noted that air emissions addressed by the labels include not only the pollutants that contribute to global warming, but a broader set of substances. The first hotspot arises in the life cycle stage raw material production, which contributes to over 20% of the total impact (see Table 3). Out of seven analysed labels, only the Blue Angel Textiles sets specific requirements on air emissions for the raw material production. The latter include thresholds for sulphur compound emissions, volatile organic compounds (VOCs), and nitrogen oxides. Air emissions in the textile manufacturing phase contribute to over one-third of the total impact. This hotspot is addressed by all analysed labels (except CmiA) using specific requirements. In contrast, air emissions in the use stage, which according to PEFCR has around 8% of the total impact, are addressed by only one label: GOTS.

It can be summarized that only one of the hotspots identified by PEF is not covered by the selected labels: Water use in the life cycle stage garment use. Four out of the five hotspots are addressed by the Blue Angel textile label, followed by GOTS with three addressed hotspots.

5. Discussion

5.1. Focus Points and Gaps in the Textile Ecolabeling

According to the applied characterization scheme, all labels show strong similarities with regards to the analysed attributes, e.g., most labels have an international focus (except CmiA), operate mainly on the product level, and focus in particular on the end consumer (i.e., B2C). All labels have a multi-aspect approach and intend environmental excellence of the certified products. However, the scope with regard to the considered environmental aspects and life cycle stages significantly differs between the analysed labels. While the labels have a comparably similar focus with regard to toxicity and water use in the raw material production and textile manufacturing phase, other impacts (e.g., land use) and life cycle stages (e.g., distribution and use phase) are considered sporadically by different labels. Furthermore, the way the requirements set significantly differ from label to label, i.e., a label provides only general requirements, only specific requirements, or both. This can lead to large differences in the broadness and strictness of the provided requirements. For example, general requirements for cotton cultivation stage include sourcing of organic cotton. Although organic production usually leads to a reduction of fertilizers and pesticides use, it does not set any restrictions on water use (e.g., as it is done by the specific requirement set by the CmiA label). Nevertheless, cotton cultivation is usually associated with high water consumption, which remains not addressed if only a general requirement is applied for this life cycle stage. In the textile manufacturing phase, general requirements include, for example, implementation of environmental management on a company level, which may reduce environmental impacts that are not directly related to the product, but the organization as a whole (e.g., waste management). In this case, the label with both types of requirements (general and specific) has an advantage over the labels that adopt only a general or specific requirement.

It can be summarized that although the analysed labels have strong similarities (according to the characterization scheme), they are not comparable due to large differences between considered life cycle stages and environmental impacts, as well as the way the requirements are set (i.e., general or specific). These findings are similar to the results of Clancy et al. who demonstrated different scopes of six textile ecolabels with regard to considered life cycle stages [24]. The authors demonstrate that a strong focus is set on the resource acquisition/farming, production of yarn/fabric, and garment manufacturing phases, which is in line with current study. The focus on the use phase was identified for three labels, which is not confirmed for the labels analysed in the current study. Possible reasons for this are discussed later in this section.

With regard to the environmental impacts, it can be seen that the hotspots (water use and air emissions) in the raw material production and textile manufacturing phase are covered by most labels. In contrast, the hotspot related to the water use in the textile usage phase remains a gap despite its high relevance. For the use phase, only toxicity and air emissions are explicitly addressed by one label, while water is not addressed at all. Of course, it is questionable whether and how producers and consumers can influence this life cycle phase, especially explicitly. Even though some studies demonstrate that fibre and garment type can influence consumer behaviour, they also show that laundry practices are highly dependent on cultural and country specific effects (habit of hand washing, quality of washing machines, use of tumble dryers) [61,62]. They are further linked to garment use, social auditing, cultural norms, garment aesthetics, life stage, and household arrangements [63]. The extent to which producers and consumers can influence laundry practices is therefore complex, and further research is needed to identify which requirements can sufficiently influence the impacts associated with the use phase of textiles. Therefore, although the labels Blue Angel Textile and GOTS provide some criteria for the use phase (e.g., the tolerance of change in dimensions during washing and drying or (colour) fastness to washing, perspiration, rubbing, light, and salvia) they were not considered for the evaluation of the use phase in this study. The analysed labels therefore leave out some crucial environmental aspects and life cycle steps, especially in the downstream life cycle stages. For this reason, the claim that the textiles sealed with one of the analysed labels are produced in an environmentally friendly manner can only be partly confirmed.

The ecolabels' function, as defined in chapter 1.2, is to fill a gap in the consumer's knowledge about environmental product information that the consumer cannot obtain on their own [44]. The ISO norm 14,020 further claims that an ecolabel shall consider the life cycle of a product or service from production to final deposit [43]. The fact that the distribution, garment use, and garment disposal steps are neglected by the analysed ecolabels shows that this is not necessarily the case. It is therefore questionable if these ecolabels successfully fill the environmental information gap as they ought to.

One solution to increase comprehensibility given the large number of different focus areas of the ecolabels is an umbrella ecolabel. The idea of such an umbrella ecolabel is to form one ecolabel that represents compliance with many different ecolabels, each with different focus areas, so that the umbrella label addresses the sum of important aspects. Consumers can then rely on this umbrella label, instead of familiarizing with various individual ecolabels. One such umbrella label is The Grüner Knopf, which has been developed by the German Federal Ministry for Economic Cooperation and Development and was introduced in September 2019 in Germany [64]. The Grüner Knopf is based on recognized ecolabels in the areas of social and environmental sustainability. For environmental sustainability, so far, nine ecolabels are named that qualify as a basis for the Grüner Knopf. Out of those nine ecolabels, four were analysed in this article: GOTS, Blue Angel, bluesign®, and Cradle to Cradle Certified™ (silver). The requirements for environmental sustainability are set by the Grüner Knopf in the areas of waste water, air emissions, chemical residues, chemicals harmful to health, chemicals harmful to the environment, EU Chemicals Regulation REACH, biodegradability, use of natural fibres, and use of synthetic fibres [64]. These requirements focus only on textile production, leaving out all other life cycle steps. The Grüner Knopf ecolabel therefore, so far, does not add to the existing ecolabels when it comes to environmental sustainability, or the needed informational value.

5.2. Limitations of Results

Seven labels with different scopes were selected for the evaluation. Although the selected labels are broadly applied in textile sector, they cannot be seen as a representation of all existing textile ecolabels.

According to the analysed life cycle steps and environmental aspects, five focus areas (toxicity, water use, air emissions, land use, recycling) were identified. However, the analysis solely considers whether these focus areas are addressed by the labels, but does not evaluate how strict the criteria are. For example, the differences between the thresholds for the emissions in water during the textile production set by different levels are not evaluated. Therefore, a quantitative comparison of the criteria adopted by different labels or definite statements on the quality of those criteria or the ecolabels themselves is impossible. The results merely present if environmental aspects are explicitly addressed in a certain life cycle step, but do not inform about the quality or quantity of the criteria. The seven ecolabels themselves are not directly comparable nor are the differently established criteria. An effort to make the criteria comparable would need to include a way to break down the different approaches and label structures. For this, an approach would be needed to make a single set of requirements comparable to a five level system as well as a traffic light system of requirements adopted by some of the evaluated labels.

The analysis of the hotspots in the textile life cycle includes only the aspects water use (consumption and pollution) and air emissions (based on the PEFCR impact category climate change). Other hotspots could not be evaluated due to missing data. Nevertheless, existing literature highlights further hotspots. As demonstrated in several studies, toxicity effects are particularly relevant in the raw material production (e.g., due to application of pesticides in the cotton cultivation [65,66] or input of chemicals during the production of man-made fibres [67] and textile manufacturing (e.g., mainly due to the input of dyes and auxiliary materials during the textile finishing) phases. While all labels (except CmiA) have a strong focus on the textile manufacturing step, for which several restrictions and thresholds with regard to the usage of toxic substances are provided, toxicity impacts in the raw material production stage are addressed only by three labels. Still, toxicity in the raw material production phase is indirectly addressed by other labels by means of the general criteria like organic cultivation and/or compliance with the legislation on the regulation of chemicals. Another relevant hotspot is land use in the raw material production, which is however addressed only by two labels: Blue Angel Textiles (requirement to source cellulose from wood cultivated according to sustainable forestry management principles) and CmiA (e.g., cutting primary forest is an exclusion criteria, further requirements are available). Production of natural fibres usually leads to the cultivation of a monoculture on large areas. This can lead to such environmental impacts as loss of biodiversity [68] or an increase of wild fires, e.g., in the case of eucalyptus forests, which are often used as a raw material for the production of cellulose fibres [69]. All these aspects are underrepresented in the requirements of the labels.

The analysis further disregards unmentioned environmental aspects. For example, microplastics pollution, which are a relevant environmental aspect, as the use of fibres based on petrochemicals is constantly increasing [8]. This affects environmental aspects such as air emissions during raw material production and emissions to water during the garment use phase. With each washing cycle, microplastics enter the ecosystem. As this specific environmental aspect was not included in any of the criteria, it was not included in the analysis even though it is a relevant aspect.

A further limitation to the results is that due to the scope of this research, it was not possible to consider the social criteria of textile production. Hence, even though some of the labels address social criteria, these were not evaluated. Including the element of social criteria makes the discussion, especially around the understandability of ecolabels for consumers and use of an umbrella ecolabel, even more complicated.

6. Conclusions

The goal of this paper was to characterize selected labels to identify their strength and weaknesses as well as to determine whether they address all relevant environmental aspects over their life cycle. The analysis showed that none of the selected labels considers all relevant life cycle phases or all relevant environmental impacts. While a clear focus is set on the upstream life cycle phases and for the environmental aspects toxicity, water use, and air emissions, significant gaps in the downstream phases could be identified. Overall, the Blue Angel Textile and the GOTS label performed best. This questions whether the ecolabels are able to fill consumers' information gaps for environmental information as well as lead to more environmental friendly consumption and products.

Based on the presented results of the analysis, several recommendations for policy and practitioners can be derived. The use phase of textiles needs to be considered, because impacts arise due to water and electricity use for washing as well as maintenance of textiles. However, impacts due to water use and electricity, which highly depend on consumer behaviour, are challenging to include in a label. Rather, a reduction of impacts should be reached by awareness rising of consumers. The detergent sector attributes impacts of water use and electricity for washing to the detergents life cycle and is carrying out awareness rising campaigns to change consumers washing behaviours for several years now. By teaming up on these awareness raising campaigns, the use phase of textiles might be reduced in the future. This aspect maintenance should be included in labels as it can be more easily measured and does not fully rely on consumer behaviour, e.g., certain companies are now offering lifelong maintenance and repairs. Further, the mandatory use of labels should be discussed. There are several reasons why ecolabels are mostly a voluntary policy instrument (e.g., costs for company and consumer). However, due to the sever impacts of the textile sector, a mandatory application of labels should be considered, similar as it is done for energy intensive products (e.g., European energy consumption labelling scheme). Different approaches are possible, e.g., deriving a mandatory European label for textile or defining clear benchmarks with regard to environmental impacts that need to be fulfilled by all companies on the European market. One option to do that could be the use of umbrella labels as they enhance not only comprehensibility, but also bring the best of different labels with regard to considered aspect and well-formulated criteria together. For the voluntary market, strengthening exiting well-performing eco labels like the German Blue Angel by carrying out information campaigns to inform more consumers about these labels, and therefore increasing the pressure for more companies to label their products. Further, all labels should be working on including unaddressed relevant environmental impacts.

Supplementary Materials: The following are available online at https://www.mdpi.com/2071-1050/13/4/1751/s1, Table S1: Textile LCA studies in the literature, Table S2: Blue Angel Textiles, Table S3: bluesign®, Table S4: Cotton made in Africa, Table S5: Cradle to Cradle Certified™, Table S6: GOTS Global Organic Content Standard, Table S7: Global Recycled Standard (GRS), Table S8: VAUDE Green Shape.

Author Contributions: F.D. and N.M. were mainly responsible for the conceptualization of the paper as well as developing the methodology. Carrying out the search for the labels and analysis of the labels was done by F.D. The draft of the paper was written by F.D., while N.M., V.B.; and M.F.

reviewed the paper draft. All authors were involved in reviewing and editing the final paper draft. All authors have read and agreed to the published version of the manuscript.

Funding: This research received no external funding.

Institutional Review Board Statement: Not applicable.

Informed Consent Statement: Not applicable.

Data Availability Statement: Data not applicable.

Conflicts of Interest: The authors declare no conflict of interest.

References

1. IPCC. Summary for Policymakers. In *Global Warming of 1.5 °C. An IPCC Special Report on the Impacts of Global Warming of 1.5 °C Above Pre-Industrial Levels and Related Global Greenhouse Gas Emission Pathways, in the Context of Strengthening the Global Response to the Threat of Climate Change, Sustainable Development, and Efforts to Eradicate Poverty*; World Meteorological Organization: Geneva, Switzerland, 2018.
2. Ripple, W.J.; Wolf, C.; Newsome, T.M.; Barnard, P.; Moomaw, W.R. World Scientists' Warning of a Climate Emergency. *BioScience* **2019**. [CrossRef]
3. Hook, L. Millions of Demonstrators Join Largest Climate Protest in History. *Financial Times* 2019. Available online: https://www.ft.com/content/d1b401d6-dbc1-11e9-8f9b-77216ebe1f17 (accessed on 20 October 2020).
4. Laville, S.; Watts, J. Across the Globe, Millions Join Biggest Climate Protest Ever. *The Guardian* 2019. Available online: https://www.theguardian.com/environment/2019/sep/21/across-the-globe-millions-join-biggest-climate-protest-ever (accessed on 22 October 2020).
5. Rodriguez, C. Biggest-Ever Climate Protest in Photos: Greta Thunberg and the World's Youth Demand Action. *Forbes* 2019. Available online: https://www.forbes.com/sites/ceciliarodriguez/2019/09/21/biggest-ever-climate-protest-in-photos-greta-thunberg-and-the-worlds-youth-demand-action/ (accessed on 30 September 2020).
6. Sengupta, S. Protesting Climate Change, Young People Take to Streets in a Global Strike. *The New York Times* 2019. Available online: https://www.nytimes.com/2019/09/20/climate/global-climate-strike.html (accessed on 30 September 2020).
7. Ellen MacArthur Foundation. *A New Textiles Economy: Redesigning Fashion's Future*; Ellen MacArthur Foundation: Cowes, UK, 2017.
8. De Falco, F.; Gullo, M.P.; Gentile, G.; Di Pace, E.; Cocca, M.; Gelabert, L.; Brouta-Agnésa, M.; Rovira, A.; Escudero, R.; Villalba, R.; et al. Evaluation of microplastic release caused by textile washing processes of synthetic fabrics. *Environ. Pollut.* **2018**, *236*, 916–925. [CrossRef]
9. Amed, I.; Balchandani, A.; Beltrami, M.; Berg, A.; Hedrich, S.; Rölkens, F. *The State of Fashion 2019: A Year of Awakening*. McKinsey, Business of Fashion. 2018. Available online: https://www.mckinsey.com/~{}/media/McKinsey/Industries/Retail/Our%20Insights/The%20State%20of%20Fashion%202019%20A%20year%20of%20awakening/The-State-of-Fashion-2019-final.ashx (accessed on 8 August 2020).
10. Thredup Share of Consumers Who Prefer Apparel from Environmentally Friendly Brands Worldwide in 2013 and 2018. Available online: https://www.statista.com/statistics/1008404/share-of-consumers-who-prefer-apparel-from-environmentally-friendly-brands-worldwide/ (accessed on 11 November 2019).
11. Gruère, G. *A Characterisation of Environmental Labelling and Information Schemes*; OECD Environment Working Papers; OECD Publishing: Paris, France, 2013; pp. 15–37.
12. Jastram, S.M.; Schneider, A.-M. Sustainable fashion governance at the example of the partnership for sustainable textiles. *UWF UmweltWirtschaftsForum* **2015**, *23*, 205–212. [CrossRef]
13. Global Organic Textile Standard. Available online: https://www.global-standard.org/ (accessed on 25 September 2019).
14. Grüner Knopf: Das Staatliche Siegel Für Nachhaltige Textilien. Available online: https://www.gruener-knopf.de/ (accessed on 10 November 2019).
15. BCI. Better Cotton Initiative. Available online: https://bettercotton.org/ (accessed on 26 June 2020).
16. Roadmap to Zero. Available online: https://www.roadmaptozero.com/about (accessed on 17 September 2020).
17. Spengler, L.; Jepsen, D.; Zimmermann, T.; Wichmann, P. Product sustainability criteria in ecolabels: A complete analysis of the Blue Angel with focus on longevity and social criteria. *Int. J. Life Cycle Assess.* **2019**, *25*, 936–946. [CrossRef]
18. Minkov, N.; Bach, V.; Finkbeiner, M. Characterization of the Cradle to Cradle Certified™ Products Program in the Context of Eco-labels and Environmental Declarations. *Sustainability* **2018**, *10*, 738. [CrossRef]
19. Bach, V.; Minkov, N.; Finkbeiner, M. Assessing the Ability of the Cradle to Cradle Certified™ Products Program to Reliably Determine the Environmental Performance of Products. *Sustainability* **2018**, *10*, 1562. [CrossRef]
20. Koszewska, M. Social and Eco-Labelling of Textile and Clothing Goods as Means of Communication and Product Differentiation. *Fibres Text. East. Eur.* **2011**, *19*, 20–26.
21. Partzsch, L.; Zander, M.; Robinson, H. Cotton certification in Sub-Saharan Africa: Promotion of environmental sustainability or greenwashing? *Glob. Environ. Chang.* **2019**, *57*, 101924. [CrossRef]

22. Targosz-Wrona, E. Ecolabelling as a Confirmation of the Application of Sustainable Materials in Textiles. *Fibres Text. East. Eur.* **2009**, *75*, 21–25.
23. Henninger, C.E. Traceability the New Eco-Label in the Slow-Fashion Industry?—Consumer Perceptions and Micro-Organisations Responses. *Sustainability* **2015**, *7*, 6011–6032. [CrossRef]
24. Clancy, G.; Fröling, M.; Peters, G. Ecolabels as drivers of clothing design. *J. Clean. Prod.* **2015**, *99*, 345–353. [CrossRef]
25. Minkov, N.; Lehmann, A.; Finkbeiner, M. The product environmental footprint communication at the crossroad: Integration into or co-existence with the European Ecolabel? *Int. J. Life Cycle Assess.* **2019**, *25*, 508–522. [CrossRef]
26. Blue Angel. Available online: https://www.blauer-engel.de/en/products/home-living/textiles (accessed on 15 October 2019).
27. Bluesign®—Solutions and Services for a Sustainable Textile. Available online: https://www.bluesign.com/en (accessed on 16 October 2019).
28. CmiA—Cotton Made in Africa. Available online: www.cottonmadeinafrica.org (accessed on 25 September 2019).
29. Cradle to Cradle Products Innovation Institute. Available online: https://www.c2ccertified.org/ (accessed on 25 September 2019).
30. Textile Exchange. Available online: https://textileexchange.org/integrity/ (accessed on 16 October 2019).
31. VAUDE CSR-Report. Available online: https://nachhaltigkeitsbericht.vaude.com/gri/news/VAUDE-erhaelt-das-neue-staatliche-Textilsiegel-Gruener-Knopf.php (accessed on 10 November 2019).
32. Hackett, T. A Comparative Life Cycle Assessment of Denim Jeans and a Cotton T-Shirt: The Production of Fast Fashion Essential Items From Cradle to Gate. Master's Thesis, University of Kentucky, Lexington, KY, USA, 2015.
33. Spoor, M. The Aral Sea Basin Crisis: Transition and Environment in Former Soviet Central Asia. *Dev. Chang.* **1998**, *29*, 409–435. [CrossRef]
34. Bardecki, M.; Kozlowski, A.; Searcy, C. Environmental Impacts in the Fashion Industry: A Life-Cycle and Stakeholder Framework. *J. Corp. Citizsh.* **2012**, *45*, 15–34.
35. Laursen, S.E.; Hansen, J.; Knudsen, H.H.; Wenzel, H.; Larsen, H.F.; Christensen, F.M. *EDIPTEX: Environmental Assessment of Textiles*; Danish Ministry of the Environment/Dansih Environmental Protection Agency: Copenhagen, Denmark, 2007.
36. Steinberger, J.K.; Friot, D.; Jolliet, O.; Erkman, S. A spatially explicit life cycle inventory of the global textile chain. *Int. J. Life Cycle Assess.* **2009**, *14*, 443–455. [CrossRef]
37. Ren, X. Development of environmental performance indicators for textile process and product. *J. Clean. Prod.* **2000**, *8*, 473–481. [CrossRef]
38. Leitão, J.; De Brito, S.; Cubico, S. Eco-Innovation Influencers: Unveiling the Role of Lean Management Principles Adoption. *Sustainability* **2019**, *11*, 2225. [CrossRef]
39. Koszewska, M. CSR Standards as a Significant Factor Differentiating Textile and Clothing Goods. *Fibres Text. East. Eur.* **2010**, *18*, 83.
40. Guedes, G.; Ferreira, F.; Urbano, L.; Marques, A.D. Corporate Social Responsibility: Competitiveness in the Context of Textile and Fashion Value Chain. *Environ. Eng. Manag. J.* **2017**, *16*, 1193–1202. [CrossRef]
41. Battaglia, M.; Testa, F.; Bianchi, L.; Iraldo, F.; Frey, M. Corporate Social Responsibility and Competitiveness within SMEs of the Fashion Industry: Evidence from Italy and France. *Sustainability* **2014**, *6*, 872–893. [CrossRef]
42. Thorisdottir, T.S.; Jóhannsdóttir, L. Corporate Social Responsibility Influencing Sustainability within the Fashion Industry. A Systematic Review. *Sustainability* **2020**, *12*, 9167. [CrossRef]
43. *ISO14020:2000. Environmental Labels and Declarations—General Principles*; ISO: Geneva, Switzerland, 2000.
44. Stø, E.; Strandbakken, P.; Scheer, D.; Rubik, F. Background: Theoretical contributions, eco-labels and environmental policy. In *The Future of Eco-Labelling*; Routledge: London, UK, 2017; pp. 16–45. [CrossRef]
45. *ISO 14024:2018. Environmental Labels and Declarations—Type I Environmental Labelling—Principles and Procedures*; ISO: Geneva, Switzerland, 2018.
46. *ISO 14021:2016. Environmental Labels and Declarations—Self-Declared Environmental Claims (Type II Environmental Labelling)*; ISO: Geneva, Switzerland, 2016.
47. *ISO 14025:2006. Environmental Labels and Declarations—Type III Environmental Declarations—Principles and Procedures*; ISO: Geneva, Switzerland, 2006.
48. Bratt, C.; Hallstedt, S.; Robèrt, K.-H.; Broman, G.; Oldmark, J. Assessment of eco-labelling criteria development from a strategic sustainability perspective. *J. Clean. Prod.* **2011**, *19*, 1631–1638. [CrossRef]
49. Minkov, N.; Lehmann, A.; Winter, L.; Finkbeiner, M. Characterization of environmental labels beyond the criteria of ISO 14020 series. *Int. J. Life Cycle Assess.* **2019**, *25*, 840–855. [CrossRef]
50. Blue Angel. The German Ecolabel Textiles DE-UZ 154. 2017. Available online: https://www.blauer-engel.de/sites/default/files/publication/be-factsheet-textiles-rz-en-web.pdf (accessed on 12 March 2020).
51. Bluesign Technologies ag Bluesign®Criteria for Bluesign®Product. 2014. Available online: https://www.bluesign.com/downloads/criteria/-2019/bluesign_criteria_for_bluesign_approved_chemical_products_and_articles_for_industrial_use_v2_0.pdf (accessed on 15 March 2020).
52. Bluesign Technologies ag Bluesign®System. 2014. Available online: https://www.bluesign.com/downloads/criteria/-2019/bluesign_system_v1_0.pdf (accessed on 15 March 2020).

53. Aid by Trade Foundation (AbTF) Cotton Made in Africa (CmiA) Criteria Matrix Version 3.1. 2015. Available online: https://cottonmadeinafrica.org/wp-content/uploads/2020/03/CmiA-Standard-Criteria-Matrix-Volume-3-1.pdf (accessed on 18 March 2020).
54. Cradle to Cradle Products Innovation Institute Cradle to Cradle CertifiedTM Product Standard Version 3.1. 2016. Available online: https://cdn.c2ccertified.org/resources/certification/standard/STD_C2CCertified_ProductStandard_V3.1_030220.pdf (accessed on 2 March 2020).
55. Global Standard gGmbH GOTS Global Organic Textile Standard Version 5.0. 2017. Available online: https://www.global-standard.org/images/resource-library/documents/standard-and-manual/GOTS-Standard_5.0_deutsch.pdf (accessed on 5 March 2020).
56. Textile Exchange Global Recycled Standard 4.0. 2017. Available online: https://textileexchange.org/wp-content/uploads/2017/06/Global-Recycled-Standard-v4.0.pdf (accessed on 1 April 2020).
57. VAUDE CSR-Report Sustainability Report. Available online: https://csr-report.vaude.com/gri-en/index.php (accessed on 25 September 2019).
58. Pesnel, S.; Payet, J. Product Environmental Footprint Category Rules (PEFCR) T-Shirts. 2019. Available online: https://ec.europa.eu/environment/eussd/smgp/pdf/PEFCR_tshirt.pdf (accessed on 2 November 2020).
59. *European Commission Guidance for the Development of Product Environmental Footprint Category Rules (PEFCRs)*; Brussels, Belgium, 2017; p. 238. Available online: https://eplca.jrc.ec.europa.eu/permalink/PEFCR_guidance_v6.3-2.pdf (accessed on 1 November 2020).
60. *2013/179/EU. Commission Recommendation of 9 April 2013 on the Use of Common Methods to Measure and Communicate the Life Cycle Environmental Performance of Products and Organisations Text with EEA Relevance*; European Union: Bruxelles, Belgium, 2013; Volume 124.
61. McQueen, R.H.; Moran, L.J.; Cunningham, C.; Hooper, P.M.; Wakefield, K.A.-M. The impact of odour on laundering behaviour: An exploratory study. *Int. J. Fash. Des. Technol. Educ.* **2019**, *13*, 20–30. [CrossRef]
62. Laitala, K.; Klepp, I.G.; Kettlewell, R.; Wiedemann, S. Laundry Care Regimes: Do the Practices of Keeping Clothes Clean Have Different Environmental Impacts Based on the Fibre Content? *Sustainability* **2020**, *12*, 7537. [CrossRef]
63. Rigby, E.D. Fashion Design and Laundry Practices: Practice-Orientated Approaches to Design for Sustainability. Ph.D. Thesis, University of the Arts London, London, UK, 2016.
64. Bundesministerium für wirtschaftliche Zusammenarbeit und Entwicklung Anforderungen Im Bereich Unternehmerische Sorgfaltspflichten Für Menschenrechte Und Umwelt in Der Lieferkette. 2019. Available online: https://www.umweltbundesamt.de/sites/default/files/medien/1410/publikationen/2019-09-03_texte_102-2019_ap_1-unternehmerische-sorgfaltspflichten.pdf (accessed on 10 July 2020).
65. Khan, M.; Mahmood, H.Z.; Damalas, C.A. Pesticide use and risk perceptions among farmers in the cotton belt of Punjab, Pakistan. *Crop. Prot.* **2015**, *67*, 184–190. [CrossRef]
66. Ahmad, A.; Shahid, M.; Khalid, S.; Zaffar, H.; Naqvi, T.; Pervez, A.; Bilal, M.; Ali, M.A.; Abbas, G.; Nasim, W. Residues of endosulfan in cotton growing area of Vehari, Pakistan: An assessment of knowledge and awareness of pesticide use and health risks. *Environ. Sci. Pollut. Res.* **2018**, *26*, 20079–20091. [CrossRef]
67. Van Der Velden, N.M.; Patel, M.K.; Vogtländer, J.G. LCA benchmarking study on textiles made of cotton, polyester, nylon, acryl, or elastane. *Int. J. Life Cycle Assess.* **2013**, *19*, 331–356. [CrossRef]
68. Baudron, F.; Corbeels, M.; Monicat, F.; Giller, K.E. Cotton expansion and biodiversity loss in African savannahs, opportunities and challenges for conservation agriculture: A review paper based on two case studies. *Biodivers. Conserv.* **2009**, *18*, 2625–2644. [CrossRef]
69. Fernandes, P.M.; Guiomar, N.; Rossa, C.G. Analysing eucalypt expansion in Portugal as a fire-regime modifier. *Sci. Total. Environ.* **2019**, *666*, 79–88. [CrossRef] [PubMed]

Article

Textiles for Circular Fashion: The Logic behind Recycling Options

Paulien Harmsen *, Michiel Scheffer and Harriette Bos

Wageningen Food and Biobased Research, P.O. Box 17, 6700 AA Wageningen, The Netherlands; michiel.scheffer@wur.nl (M.S.); harriette.bos@wur.nl (H.B.)
* Correspondence: paulien.harmsen@wur.nl

Abstract: For the textile industry to become sustainable, knowledge of the origin and production of resources is an important theme. It is expected that recycled feedstock will form a significant part of future resources to be used. Textile recycling (especially post-consumer waste) is still in its infancy and will be a major challenge in the coming years. Three fundamental problems hamper a better understanding of the developments on textile recycling: the current classification of textile fibres (natural or manufactured) does not support textile recycling, there is no standard definition of textile recycling technologies, and there is a lack of clear communication about the technological progress (by industry and brands) and benefits of textile recycling from a consumer perspective. This may hamper the much-needed further development of textile recycling. This paper presents a new fibre classification based on chemical groups and bonds that form the backbone of the polymers of which the fibres are made and that impart characteristic properties to the fibres. In addition, a new classification of textile recycling was designed based on the polymer structure of the fibres. These methods make it possible to unravel the logic and preferred recycling routes for different fibres, thereby facilitating communication on recycling. We concluded that there are good recycling options for mono-material streams within the cellulose, polyamide and polyester groups. For blended textiles, the perspective is promising for fibre blends within a single polymer group, while combinations of different polymers may pose problems in recycling.

Keywords: textile; recycling; circular fashion; polymer structure

Citation: Harmsen, P.; Scheffer, M.; Bos, H. Textiles for Circular Fashion: The Logic behind Recycling Options. *Sustainability* **2021**, *13*, 9714. https://doi.org/10.3390/su13179714

Academic Editors: Hanna de la Motte and Asa Ostlund

Received: 3 June 2021
Accepted: 3 August 2021
Published: 30 August 2021

Publisher's Note: MDPI stays neutral with regard to jurisdictional claims in published maps and institutional affiliations.

Copyright: © 2021 by the authors. Licensee MDPI, Basel, Switzerland. This article is an open access article distributed under the terms and conditions of the Creative Commons Attribution (CC BY) license (https://creativecommons.org/licenses/by/4.0/).

1. Introduction

The current textile industry uses large amounts of non-renewable resources and applies hazardous substances and polluting processes. On top of this, the ever-spreading trend of fast fashion has led to fast-fashion retailers selling clothing expected to be discarded after being worn only a few times [1]. Hence, new solutions to reduce the use of (virgin) resources need to be developed and implemented. One of the proposed routes is to increase circularity in the textiles and clothing industry [2].

However, what is the definition of circularity in relation to the textile industry? Circularity is a concept that originates from the field of industrial ecology, combined with circular design concepts such as cradle to cradle [3,4]. The underlying concerns are the ever-increasing depletion of non-renewable feedstock. Circular solutions thus aim at fulfilling societal demand while minimising the input of virgin resources. Circular concepts, therefore, go much further than just recycling materials into new products. As stated by Rosa et al., an actual transition towards a Circular Economy requires relevant changes along the whole value chain, not only for waste generation and resource use but also for adopted market strategies and business models [5].

Several circularity strategies exist to reduce the consumption of natural resources and materials and minimise waste production. They can be prioritised according to their levels of circularity, i.e., the 10R-strategy [6]. The hierarchy of circular solutions relevant for the textiles and clothing market segments can be presented in Figure 1, which was built on this 10R-strategy.

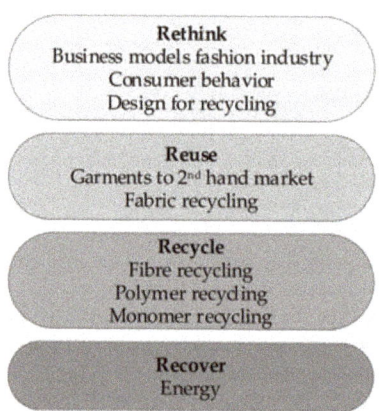

Figure 1. Different options for circular textiles based on the 10R-strategy. The higher in order, the more preferable and sustainable the options are.

Ideally, waste is prevented by changing consumer buying behaviour, wearing clothes for a longer period, implementing other business models in the fashion industry such as textile rentals or by applying design for longevity principles (Rethink). Ideally, when a garment is discarded, it is reused by another customer through the second-hand market. When the garment is no longer wearable, it can be converted into a product of lower value, such as wiping rags (Reuse). When the fabric is no longer usable as such, recycling techniques come into play. Here, a distinction can be made between fibre, polymer or monomer recycling (Recycle). When recycling is also no longer possible, recovery of energy (Recover) is the final option.

In this publication, we focus mainly on the Recycle part. According to Directive 2008/98EC [7], recycling means 'any recovery operation by which waste materials are reprocessed into products, materials or substances, whether for the original or other 'purposes' [8]. The emphasis in our publication is on the options for recycling textile fibres back into textiles, but other end-uses of recycled fibres are also mentioned.

Initiatives and technologies for textile recycling are emerging rapidly, and similarities with plastic recycling are apparent. However, for textiles, the large variety of materials used (i.e., fibres, auxiliaries) combined with a high level of structural complexity is substantially different. For a better understanding of developments in textile recycling, we encountered three fundamental problems.

The first is the definition of the classes of textile fibres. The various fibre types are classed as either being natural (vegetable, animal, mineral) or artificial (regenerated/natural, synthetic and inorganic) [9,10]. However, it is much wiser for recycling purposes to classify the fibres based on their main chemical bonds instead of their origin, as fibres with the same kind of chemical bonds usually have similar chemical and, often, physical characteristics. In this publication, we propose a new classification of textile fibres that matches recycling technologies.

A second issue is the classification of recycling technologies. There is no standard definition, and many descriptions are used for textile recycling [11]. Sandin and Peters propose a "topology of textile reuse and recycling" based on the level of disassembly of the recovered material (fabric, fibre, polymer/oligomer and monomer recycling) [12]. They state that the "systematisation of recycling routes into mechanical, chemical and thermal ones is ambiguous and questionable". We support this statement, but at the same time, we argue that the classification based on disassembly is also not entirely suitable. In this paper, we combine the two approaches into one new classification of textile recycling based on the level of disassembly combined with mechanical, physical and chemical recycling methods. This approach is based on the molecular structure and structure/properties relation of the polymers that make up the fibres, enabling us to unravel the logical and preferred recycling routes for the various types of textile fibres.

The last problem we face is the lack of clear communication about (1) the technological progress by companies and brands and (2) the benefits of textile recycling from a consumer perspective. Wagner and Heinzel state that "consumers' awareness, attributed importance, and perceived value become crucial for the success of recycled synthetic textile products" [11]. The choice between virgin and recycled textile products depends on the perceived benefits or disadvantages. Recycled content is often thought to be synonymous with low quality, but we show that the final quality strongly depends on the type of fibre and the technology used.

The main aim and novelty of this work is the design of improved classification systems for textile fibres and textile recycling. Our hypothesis is that an approach based on polymeric structures contributes to a better understanding of textile recycling and a better definition of the challenges ahead.

We explain our applied methods in Section 2. In Section 3, our main results are presented, i.e., a new classification system of textile fibres and textile recycling, with the corresponding recycling methods (mechanical, physical and chemical). This section is complemented with a schematic overview of resources and processes to produce garments in a Linear Fashion manner based on virgin resources or in a Circular Fashion manner with recycling and reuse. Section 4 is dedicated to discussions on recycling options for the main polymer types defined, both from a theoretical approach and based on commercial activities published on websites. Based on this discussion, a feasibility assessment was created, followed by a discussion on the challenge of blended textiles. Finally, Section 5 provides the limitations of our study, some concluding remarks and suggestions for further actions to be taken.

2. Methods

A new classification system for textile fibres that combines well with the various recycling methods available was designed. This classification system was set up by first determining the primary textile fibres from Mather [9], their corresponding polymeric structures and main chemical linkages. This resulted in six polymer types that are reviewed throughout this publication: cellulose, polyester, polyamide, polyurethane, polyolefin and polyacrylic. The following step was placing the textile fibres in their corresponding polymer category. As a final step, the type of polymer (a natural, addition or condensation polymer) was also included, which is relevant for recycling.

In addition, a new classification system for textile recycling was designed. We adopted a systematisation of recycling routes presented by the Ellen MacArthur Foundation [2] and Sandin and Peters [12], which is based on the level of disassembly: recycling to fibre, polymer or monomer. Each of these recycling routes requires (several) subsequent recycling methods to arrive at the desired product. With the polymeric structure of the fibres in mind, we distinguished sorting, mechanical, physical and chemical methods. As a final step, the classification of textile recycling was coupled with the appropriate recycling methods.

When the linear fashion chain transforms into a circular fashion chain, resources will partly change from virgin to recycled content. How this recycled content changes the textile supply chain is illustrated by one schematic. To this end, the key stages in the apparel life cycle were gathered from Tomaney [13], and linked to the output streams from recycling (on the level of fibres, polymers and monomers).

The discussion section consists of three parts. The applicability of the three recycling routes (fibre, polymer, monomer) was determined for the six polymer types. The physical structure/property relations and chemical options to break down the primary polymeric bonds were determined by desk research. The polymeric structure of the fibres and the main bonds in the polymer backbone served as a guideline to determine which methods are potentially applicable and which are not realistically possible. The potentially available methods were illustrated by the recycling initiatives that resulted from our (scientific and grey) literature review.

Based on these results and our own judgement, we constructed a feasibility assessment for the main textile polymers and fibres, emphasising options for textile to textile. For this assessment, we assumed that all textiles consist of only one type of fibre (monomaterial textiles).

However, textiles are often not composed of one type of fibre but of a variety of fibres, i.e., blended textiles. The recycling classification system and feasibility assessment were used to explore the recycling options for these blended textiles. An estimation for common textile blends was presented, including an outline of current recycling initiatives in this field.

3. Results

3.1. Classification of Textile Fibres

Textiles mainly consist of fibres, i.e., very long (2–5 cm) and very thin structures. Landi [14] describes fibre as a "unit matter with a length at least 100 times its diameter, a structure of long-chain molecules having a fixed preferred orientation, a diameter of 10–200 microns, and flexibility". Thus, all fibres have a molecular structure that contributes to their specific attributes and properties. All fibres are characterised by the common characteristics of small diameter relative to its length, flexibility and fineness [10].

Classification of textile fibres can be done in many ways. A widely used method to classify fibres is based on their origin and resources, i.e., animal, vegetable, regenerated or synthetic (mainly fossil-based) fibres. Although the origin of textile fibres is an important parameter, we propose a new classification for the definition of recycling options and the circular use of textiles. This classification is based on chemical groups and bonds that form the backbone of the polymers that the fibres are made of and that impart characteristic properties to the fibres. For recycling purposes, it is useful to classify the various types of textile fibres based on their main chemical bonds, as fibres with the same chemical bonds usually have similar chemical and often physical characteristics. Regarding textile recycling, this classification method is preferred, as it helps in evaluating the options that are available in preventing the generation of textile waste. Table 1 shows the main polymer groups discussed in this publication, i.e., polysaccharides (i.e., cellulose), polyesters, polyamides, polyurethanes, polyolefins and polyacrylics, and presents the chemical links in the polymeric backbone that are relevant for recycling approaches.

Table 1. Classification of textile fibres based on polymer linkages.

Polymer	Polysaccharides: Cellulose	Polyester	Polyamide	Polyurethane	Polyolefin	Polyacrylic
Essential linkage [1]	β-glycosidic	ester C—O—C	amide C—N—C	urethane C—O—C—N—C	alkane C—C	acrylonitrile C—C, C≡N
Fibre examples	Cotton (natural) Linen (natural) Viscose (natural) Lyocell (natural)	PET [2] (condensation)	Wool (natural) Silk (natural) Nylon (condensation)	Elastane (condensation)	PP [2] (addition) PE [2] (addition)	Acryl (addition) Modacryl (addition)
Melting point	No	Yes	No (natural) Yes (condensation)	Yes	Yes	No

[1] Only for cellulose is this the repeating unit; for the other polymers, it denotes the linkage relevant for recycling. [2] PET: polyethylene terephthalate, PP: polypropylene, PE: polyethylene.

Polymers can also be categorised as being natural, addition or condensation polymers. For the main textile fibres, this is also included in Table 1. Natural polymers are of biological origin and are formed by living organisms such as plants or animals. Cellulose (polysaccharide) and protein (polyamide) are examples of natural polymers relevant for textiles. Addition and condensation polymers are synthetic polymers built up of monomeric building blocks. Condensation polymers are formed by the reaction of two different functional groups, usually originating from two or more different monomers. During the polymerisation reaction, small molecules such as water are often eliminated. Examples include polyesters, polyamides and polyurethanes. Addition polymers are formed mainly by one type of monomer. The polymerisation reactions are chain reactions, with free radicals or ionic groups responsible for propagating the chain reaction. Examples include polyolefins and polyacrylics [15].

3.2. Categorising Textile Recycling and Recycling Methods

Textile recycling includes all processes on the level of fibres, polymers or monomers. Whether a garment is suitable for fibre, polymer or monomer recycling is determined in large part by the fibre composition and the chemical structure of the polymers that make up the fibres.

- Fibre recycling implies the preservation of the fibres after the disintegration of the fabric.
- Polymer recycling includes the disassembly of the fibres while the polymers remain intact.
- Monomer recycling implies that fibres and polymers are broken down into their chemical building blocks.

Each of these recycling methods requires (several) subsequent recycling methods to come to the desired product. We distinguish sorting, mechanical, physical and chemical methods. Because stakeholders do not always interpret terms such as mechanical and chemical in the same way, we propose the following definitions that help us present the different recycling methods for textiles in an orderly and recognisable fashion:

- Mechanical methods break down the fabric and retain the fibres by cutting, tearing, shredding or carding. The fibre length is reduced as an unwanted side effect, thereby affecting the spinnability and yarn strength [16]. The fibres will have a shorter fibre length than the original fibres, and some dust will be generated [17].
- Physical methods use physical processes to make the fibres or polymers suitable for reprocessing, either by melting or dissolving them. With physical recycling, the structure of the fibres is changed, but the polymer molecules that make up the fibres remain intact. After melting or dissolving, either melt spinning or solution spinning can be used to form a new filament (i.e., a fibre of infinite length).
- Chemical methods exploit chemical processes to break down fibres and polymers. The polymers that make up the fibres are either modified or broken down, sometimes to their original monomeric building blocks. This can be done by chemical or biological methods (e.g., with enzymes). After chemical recycling, the building blocks can be repolymerised into a new polymer.

In Table 2, the classification of textile recycling is coupled to the corresponding methods. It is evident that for all recycling initiatives, extensive sorting is needed, as mixtures of different materials and colours will result in recycled feedstock of poor quality. Mechanical methods are often required; for fibre recycling, this is the only step, but for polymer and monomer recycling, it is a treatment prior to physical or chemical recycling methods.

Table 2. Classification of textile recycling and corresponding recycling methods.

Classification of Textile Recycling	Recycling Methods	
Fibre recycling	Mechanical methods	
Polymer recycling	Mechanical methods	Physical methods
Monomer recycling	Mechanical methods	Chemical methods

3.3. Resources, Production Methods and Recycling Routes in the Linear and Circular Textile Industry

A schematic overview of the relationship between resources, intermediate products and various processes applied in the textile industry is presented in Figure 2. Several subsequent standard processing steps are needed to make a garment from renewable or fossil resources.

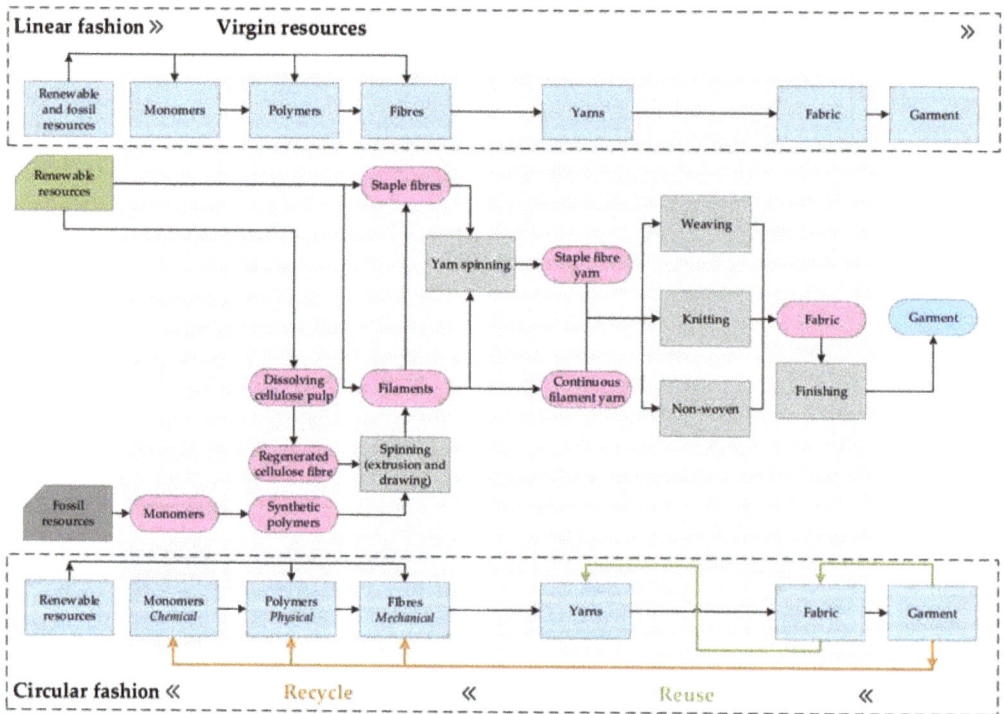

Figure 2. Relation between resources and processes to produce garments. Above is the linear route, below is the circular route.

In the current linear fashion chain, virgin resources can be either of renewable origin, coming from plants or animals, or fossil-based (Figure 2, upper and middle part). Fossil resources always enter the scheme on the monomer level, as the resulting synthetic fibres are built-up of monomeric building blocks that are polymerised into polymers. Synthetic polymers are spun into a fibre shape by extrusion spinning, either from the melt or the solution, and then spun into a yarn. Renewable resources enter the scheme on the polymer or fibre level, not yet on the monomer level. In a future scenario, this may change, as it is also possible to make synthetic polymers from renewable resources. On the polymer level, we find the natural polymer cellulose, mainly harvested from wood to produce cellulose pulp. This cellulose is subsequently spun into a fibrous shape by extrusion spinning, after which regenerated cellulose fibres are formed (continuous filaments), e.g., viscose or lyocell. On the fibre level, we find natural fibres that can be staple fibres (cotton, wool, linen) or filaments (silk), and they can be yarn spun after minor processing.

In the circular fashion chain, resources come from renewable resources and garments (post-consumer) or residues produced in the factories in the form of yarns or fabrics (post-industrial) [18] (Figure 2, lower part). These post-industrial and post-consumer residues are not yet optimised for recycling. There are distinct differences between these streams. It is widely recognised that the post-consumer stream forms the most challenging one (not well-defined, contaminated, decreased fibre quality due to washing and wearing). In addition to the recycled input, virgin renewable resources will always be required to overcome quality loss when polymer and fibre recycle streams are applied, in contrast to the recycling route via monomers, as (re)starting from the monomer level will always result in virgin quality fibres.

As shown in Figure 2, monomer, polymer and fibre recycling each follow a specific pathway and re-enter the textile production cycle at a different level. The preferred entrance level is highly dependent on the type of material that needs to be recycled. The overall aim is to keep the structure of the materials intact as much as possible and to minimise processing.

4. Discussion

4.1. Recycling Methods for Six Polymer Types: Theoretical Approach and Commercial Activities

This section presents the applicability of the three different recycling approaches for the six polymer types: cellulose, polyester, polyamide, polyurethane, polyolefins and polyacrylics.

4.1.1. Cellulose

Only two routes are available for cellulose textiles when new textile fibres are envisioned as the end-product: mechanical recycling or physical recycling. A third option, the breakdown of cellulose to glucose molecules by chemical or enzymatic methods, is currently less relevant for textiles.

- Mechanical recycling to fibres Table 3 shows that several mechanical cellulose recycling factories produce a variety of products and use cotton as input. Mechanical recycling of cotton yields fibres typically applied in the nonwoven industry and as flock (very small fibres used to create texture on surfaces). Recycled cotton fibres are shorter than virgin ones and thus more difficult to spin. Mechanically recycled fibres are often mixed with (longer) virgin fibres such as PET or cotton for woven applications. About 95% of the recovered fibres by the mechanical recycling of cotton are directly processed into nonwovens for the automotive industry, appliances, drainage systems and geotextiles [17].
- Physical recycling to polymers Polymer recycling of cellulose materials to yield regenerated cellulose fibres is an excellent option to recover and reuse cellulose polymers from residue streams. The resource to produce regenerated cellulose is usually wood, but other cellulose-containing resources could also be used, for example, post-consumer textiles. Possible bottlenecks are the contamination of the garments with other types of fibres and the presence of finishing agents and dyes. Cotton fibres are almost entirely made up of cellulose. They are excellent candidates as feedstock for the viscose and lyocell process, and this is applied on a small scale by several parties.

Table 3. Cellulose recycling initiatives (partly adapted from [19]).

Company	Input Stream	Product	Status	Ref.
Mechanical recycling to fibres				
Frankenhuis	Post-consumer textile waste (cotton)	Textile fibres for nonwovens	Commercial, the Netherlands	[20]
Wolkat	Post-consumer textile waste (cotton)	Yarns for textiles and nonwovens	Commercial, the Netherlands and Morocco	[21]
Belda Llorens	Cotton waste a.o.	EcoLife yarns, blended with virgin, for textiles	Commercial, Spain	[22]
Geetanjali Woollens	Cotton waste a.o.	Recycled cotton fibre and yarn, blended with virgin for textiles	Commercial, India	[23]
Ferre	Cotton waste	Recover yarns, blended with virgin for textiles	Commercial, Spain	[24]
Velener Textil GmbH	Post-industrial cotton yarns	WECYCLED® cotton fibres, blended with virgin for textiles	Commercial, Germany	[25]
Physical recycling to polymers				
Lenzing AG with Refibra™ (lyocell)	20% post-industrial cotton scraps combined with 80% virgin cellulose pulp from wood	Tencel™ lyocell fibres (regenerated cellulose fibres)	Commercial, Austria	[26]
Asahi Kasei	100% cotton linter, post-industrial residue of cotton processing	Bemberg™ cupro fibre (regenerated cellulose fibres)	Commercial, Japan, 17.000 mt/y	[27]
Renewcell	High-content cellulose waste (cotton, regenerated cellulose)	Dissolving pulp (Circulose pulp)	Demonstration, Sweden, 7000 mt/y	[28]
Evrnu with NuCycle™	High-content cellulose waste (cotton, regenerated cellulose)	Regenerated cellulose fibres	In development, USA	[29]
Infinited Fibre	High-content cellulose waste (cotton, regenerated cellulose)	Regenerated cellulose fibres	In development, Finland	[30]
Aalto University with Ioncell™	High-content cellulose waste (cotton, regenerated cellulose)	Ioncell™ fibres (regenerated cellulose fibres)	In development, Finland	[31]
SaXcell with SaXcell® (lyocell)	High-content cellulose waste (cotton, regenerated cellulose)	SaXcell® fibres (regenerated cellulose fibres)	In development, the Netherlands	[32]

4.1.2. Polyester

The term 'polyester' is a generic name for all polymers containing ester linkages in their polymeric chain, but in the apparel sector, 'polyester' stands for one of these types of materials: polyethylene terephthalate (PET). The recycling of PET has become increasingly important with the increasing use of PET in textiles. For textiles, PET is most suited for physical and chemical recycling, and mechanical recycling is less achievable.

- Physical recycling to polymers PET is a thermoplastic material; it melts at elevated temperature (>260 °C) and can be re-spun into fibres again. Contaminations may pose a problem in these kinds of processes, and this is therefore the reason that recycled fibres for textiles are often produced from transparent bottles (Table 4). Recycling bottles into fibres is implemented across the globe. Post-consumer transparent bottles are relatively clean and result in a high-quality rPET (recycled PET), suitable for yarn production. The use of less pure post-consumer PET items such as coloured bottles, trays and films and PET recovered from the ocean and textiles [8,33] is more challenging. If these sources cannot be sufficiently cleaned, it may be necessary to use chemical recycling methods.
- Chemical recycling to monomers and oligomers Chemical recycling methods are well suited for the production of PET fibres (Table 4). During chemical recycling, the polyester molecules are broken down into smaller fragments. These smaller fragments are recovered by separation processes such as filtration, precipitation, centrifugation

and crystallisation [34], making the removal of contaminants easier than in mechanical and physical recycling.

The technology of PET degradation or depolymerisation is done by solvolysis, i.e., reaction in a solvent in which one of the reactants is the solvent molecule. Solvolysis can be divided into various techniques, common for the depolymerisation of polycondensation polymers such as polyesters, polyurethanes and polyamides [35].

Table 4. Polyester recycling initiatives (partly adapted from [19]).

Company	Input Stream	Product	Status	Ref.
Physical recycling to polymers				
Velener Textil GmbH	PET bottles	WETURNED® PET-woven fabric	Commercial, Germany	[36]
Cumapol with CuRe Technology	Coloured PET from various sources	Transparent PET granulate	In development, the Netherlands	[37]
Chemical recycling to monomers and oligomers				
Ioniqa with glycolysis	(Coloured) PET from various packaging materials	BHET and rPET	Commercial, the Netherlands, 10 ktons/y	[38]
Jeplan with glycolysis	(Coloured) PET from various packaging materials	BHET	Commercial, Japan	[39]
Teijin	Bottles and other PET materials	DMT to ECOPET™ filament yarns	Commercial, Japan	[40]
Eastman with chemical recycling (PRT)	Polyester	Unknown	Commercial, USA	[41]
Ambercycle with enzymatic hydrolysis	Post-consumer textile waste	Cycora™ yarn for new textiles	In development, USA	[42]
Carbios with enzymatic hydrolysis	PET	Monomers EG and TPA	In development, France	[43]
Gr3n with microwave radiation	PET	Monomers	In development, Switzerland	[44]

The production of building blocks by chemical PET recycling can result in the original monomers: ethylene glycol (EG) and terephthalic acid (TPA) [35]. Other processes result in oligomers, i.e., dimers or trimers of the original building blocks, such as bis(2-hydroxyethyl), terephthalate (BHET) or dimethyl terephthalate (DMT) [34].

The most important advantage is that virgin quality PET can be achieved; the disadvantage is that chemical recycling is more expensive and requires large scale production to become economically feasible [34]. Various companies are upscaling and validating their technologies.

4.1.3. Polyamide

In contrast to the other polymer categories, the two main polyamide fibre groups, wool (natural fibre) and nylon (synthetic fibre), have very different recycling options, even though they both have amide bonds in their polymer backbone. Wool can be recycled mechanically [45], whereas for nylon, being a polycondensation polymer and a thermoplastic material, similar recycling options as for polyesters are viable [35] (Table 5).

- Wool: Mechanical recycling to fibres Recycled wool has a long tradition and is the single example of successful mechanical recycling of post-consumer textiles to new yarns for textiles. The mechanical recycling of wool is the only option for the time being. Woolen fabrics are predominantly made from long fibres and handled with care, making mechanical recycling into fibres achievable. The mechanical recycling of wool is done by similar steps as mechanical recycling of cotton. Sorted fabrics are cleaned and turned into fibres by cutting and tearing. Post-industrial wool processing residues (fibre, yarn and fabric generated during production) are routinely recycled back into the manufacturing process flow for reasons of economic efficiency [45]. Mechanical recycling of post-consumer streams is also feasible. The applicability of post-consumer residues is partly dependent upon effective ways to minimise fibre breakage and maximise residual fibre length after the mechanical pulling process.
- Nylon: Physical recycling to polymers Nylon can be melted and reshaped into new fibres of appropriate length and strength. Nylon waste is available as carpets and fishing nets (composed of nylon 6), and these materials are an important feedstock for physical nylon recycling. Recycling fishing nets is technologically feasible, and recycled nylon 6 shows similar characteristics compared to commercial nylons [46]. Recycling carpets is also possible by blending various carpet components through reactive extrusion and compatibilisation, yielding lower-quality products [47].
- Nylon: Chemical recycling to monomers For nylons, chemical recycling methods are well suited. Nylon 6 is depolymerised to retrieve its original building block caprolactam. The process design is suitable for processing 'contaminated' materials, as the product exits at the top of the reactor and residues remain on the bottom [47]. According to DSM [48], this process is not more expensive than the production of virgin caprolactam, and it is much more environmentally benign.

Table 5. Polyamide recycling initiatives (partly adapted from [19]).

Company	Input Stream	Product	Status	Ref.
Wool: mechanical recycling to fibres				
Cardato	Post-industrial Post-consumer	Yarns, "Cardato Recycled" Brand	Commercial, 22.000 mt/y, Italy (Prato)	[19]
Boer Group	Post-consumer	Yarns	Commercial, Netherlands	[49]
Geetanjali Woollens	Unknown	Recycled sheep wool and cashmere	Commercial, India	[23]
Novetex (Billie System)	Various	Unknown	Commercial, Hong Kong	[50]
Nylon: physical recycling to polymers				
Fulgar with the MSC process	Post-industrial waste	Q-Nova®, 50/50 regenerated/virgin nylon 6,6 fibre	Unknown	[51]
Nylon: chemical recycling to monomers				
Aquafil with Econyl® technology (depolymerisation to caprolactam)	Nylon 6 fishing nets, carpets, post-industrial textiles	Econyl® yarn, nylon 6	Commercial, Italy	[52]

4.1.4. Polyurethane

Recycling elastane, being a polyurethane and polycondensation polymer, is a challenge. At present, no methods are available on a pilot or demo scale. Elastane is usually present in (women's) fashion in only a very small fraction, making recycling less interesting. Elastane is usually not recycled but can be removed from the fabric to facilitate the recycling of other fibres, e.g., by solvolysis methods suitable for polycondensation polymers [35,53].

4.1.5. Polyolefin

Polyolefins such as polyethylene (PE) or polypropylene (PP) are thermoplastics and can, in theory, be physically recycled by melting and re-spinning to new fibres and yarns, but for apparel textiles, we did not find any examples. Polyolefins are addition polymers and can therefore not be recycled by chemical means, e.g., depolymerised to their monomers by solvolysis methods, as the chemical bonds in the polymer cannot be broken by these methods. These polymers can be degraded through a free radical mechanism at high temperatures, but this does not result in the formation of re-usable monomers. Instead, heterogeneous mixtures of gasses, liquids and tar are produced [54] that may be used as input for the chemical cracking process and thus add to the production of renewable bulk chemicals.

4.1.6. Polyacrylic

Polyacrylics can be mechanically recycled, comparable to wool. The process involves colour sorting, cleaning, unravelling and spinning again [49]. Polyacrylics cannot be melted, and although they can be dissolved, presumably no physical recycling methods are under development. Polyacrylics are formed by addition polymerisation and can therefore not be depolymerised by solvolysis methods.

4.2. Feasibility Assessment for Main Textile Polymers and Fibres

Recycling options for each type of textile fibre is a complex topic, as each fibre has its own optimal recycling strategy. A feasibility assessment of various recycling routes for the six polymers and the main textile fibres discussed is shown in Table 6. The emphasis here is on options for recycling textile fibres back into textile fibres.

Table 6. Most achievable (green), less achievable (orange) and not achievable (red) recycling options for main polymer groups.

Classification of Textile Recycling	Cellulose		Polyamide		Polyester	Polyurethane	Polyolefin	Polyacrylic
	Natural				Condensation		Addition	
	Cotton, Linen	Viscose, Lyocell	Wool	Nylon	PET	Elastane	PP, PE	Acrylics
Fibre recycling	green	orange	green	orange	green/orange	red	orange	green
Polymer recycling	green *	red	red	green	green	orange	orange	orange
Monomer recycling	orange	orange	green	green	green	orange	red	red

* To another type of fibre.

For natural polymers, we see multiple options for fibre and polymer recycling. The fibre recycling of natural staple fibres such as cotton is possible by mechanical means, provided that the resulting fibre length is sufficient to make yarn spinning possible. The polymer recycling of cellulose by physical means is possible by processes where the cellulose is dissolved and spun into new textile fibres. For wool, the recycling of fibres by mechanical means is the only option.

Condensation polymers such as PET and nylon can be recycled by physical and chemical methods. Polymers can be recycled by melting the material at high temperatures and re-spinning it into filaments. When high-quality products are desired, only highly purified materials can be used. Another viable option is recycling the monomers by chemical means. The input for these processes can be less pure material while the end-product is of virgin quality, but the monomers need to be 100% pure before repolymerisation can take place. Elastane has limited options for recycling.

In the category of addition polymers, the options are limited. For polyacrylics, recycling the fibres by mechanical means might be an option. The fibres can be spun into yarns to produce knitwear.

4.3. The Challenge of Blended Textiles

Contemporary textiles are increasingly composed of a variety of fibres, as single-fibre textiles often do not fulfil the requirements of today's fashion. Fibre blending, combining two or more fibre types to form a new yarn or fabric, can combine the best qualities of each fibre. In this way, the functionality of the fabric is improved [10]. A well-known example is the presence of a few per cent of elastane in jeans to improve wearing comfort [55]. Another example is a blend of PET and cotton to combine strength and comfort and to reduce the price, as polyester fibres are cheaper than cotton fibres [56]. Another reason can be to improve ease of processing, as synthetic fibres can be produced in a broader range of fibre lengths than the relatively short length of natural staple fibres, making spinning easier [56].

A wide range of fibre combinations is used nowadays. However, reliable data are hard to obtain and are mostly limited to analyses of post-consumer textiles, for example, through infrared technology. For sustainability and end-of-life options, fibre blending might not be such a good idea. In this section, we explore recycling options for blended fabrics and garments.

4.3.1. Post-Industrial and Post-Consumer Waste Streams

The most significant issues with blended textiles relate to post-consumer waste streams in the form of garments. These garments often consist of multi-material fibre compositions, making recycling very complicated. These different types of fibres need to be separated for recycling, which is difficult or even impossible [8]. Post-industrial waste streams are often well-defined and more homogeneous, as they can be collected as fibres, yarns or fabrics, thus well before the garment is produced. Post-industrial sources, therefore, have a much greater chance of achieving cost-effective, high-quality recycling than post-consumer streams [57], and some of these recycling methods are already implemented. However, according to the Ellen MacArthur Foundation, less than 1% of the material used to produce clothing is recycled into new clothing. This includes recycling after use, as well as recycling of factory offcuts [2].

All post-consumer textiles are collected and sorted in the ideal situation. However, in reality, this amount is never reached. For the Dutch situation, around 45% (136 kton) is collected separately, whereas 169 kton ends up as residual waste [58]. The amount of discarded textile per capita increased from 4.2 to 4.7 kg/y over the past 10 years [59]. Approximately half of the collected textiles are suitable for the second-hand market (mostly abroad), and the remainder is only suitable for recycling. Due to the rise of fast fashion and the decline of the second-hand market, the amount of textile suitable only for recycling (non-wearable fraction) increases every year [59].

The composition of this non-wearable fraction is not well-defined. Sorting companies such as *Leger des Heils* (Salvation Army) and Sympany sort by hand into main categories such as denim, white cotton, coloured cotton and sweaters [60]. Advanced sorting technologies such as the Fibresort [61] can sort textiles based on composition (e.g., wool, cotton, nylon, PET) and even colour but can only process mono-materials. Although we cannot support this with data, we assume that single-fibre garments are outnumbered by blended textiles, creating a big problem for textile recycling. Converting post-consumer mono-material textile fibres into new textiles is possible on a limited scale. As can be concluded from Tables 3–5, the options are even more limited for blended textiles. Table 7 shows initiatives by companies and start-ups that work on the challenge of recycling blended textiles. Most of these initiatives deal with an input stream of cotton and PET and produce a stream of PET monomers and regenerated cellulose fibres.

Table 7. Initiatives of recycling of blended textiles (partly adapted from [19]).

Company	Input Stream	Product	Status	Ref.
HKRITA/H&M Foundation with hydrothermal method	Cotton/PET	New textile fibres	Pre-industrial size facility in Hong-Kong	[62]
Worn Again with dissolution of PET and cellulose	Cotton/PET	Cellulose pulp for regenerated cellulose fibre and PET	Industrial demonstration plant to be launched in 2021	[63]
Blend Re:Wind (Rise and Chalmers University)	Cotton/PET	Viscose filaments and PET monomers	Sweden, status unknown	[64]
Sodra with Once More	Cotton/PET	Cellulose pulp for textile	Sweden, commercial	[65]
Tyton Biosciences	Cotton, PET, polycotton	Cellulose and PET monomers	USA, status unknown	[66]
Block Texx	Cotton/PET	Cellulose pulp and PET	Australia, status unknown	[67]
Resyntex with monomer recycling	Cotton, nylon, PET, wool	Monomers (glucose, TPA, EG) protein hydrolysates, polyamide oligomers	Closed EU-project	[68]
Trash2Cash with polymer recycling	Blended textile and paper waste	Regenerated cellulose	Closed EU-project	[69]

4.3.2. Common Textile Blends

Although the composition of post-consumer textiles is not well-defined, based on our experience, some combinations are frequently used. In this paragraph, the main fibre blends are discussed for each group of polymers, including possible recycling strategies.

A common blend is a cotton with a few per cent of elastane. Mechanical recycling of cotton would be the most viable option, but elastane may cause problems during mechanical processing. The cotton fibre may also be of inferior quality after use, making the conversion of cotton into regenerated cellulose fibre a better option. The best-case scenario is when the elastane fraction ends up in a solid residue while cellulose is dissolved and spun into a new fibre. The same holds for blends of elastane with other cellulose fibres such as linen, viscose or lyocell. Whether this is feasible depends on the process applied, and further investigations are needed.

A blend of cotton and PET is often used for workwear, and production volumes are high. Several approaches can be followed to deal with polycotton waste streams [70]. In general, approaches where the cellulose remains intact as much as possible are preferred, as only PET monomers can be polymerised into a new polymer. The challenge is to use mild conditions to preserve the cellulose fraction, as most of the solvolysis methods of PET will certainly degrade cellulose into smaller polymer fractions.

PET is blended with almost all other textile fibres mentioned in this publication. Although PET recycling from bottles is well underway, recycling blended textile fibres is less developed. Polymer recycling by melting and spinning into new textile fibres is less achievable due to contaminations, leaving monomer recycling as the only option. The results from a study by GreenBlue indicate that, for the chemical recycling technologies being evaluated, a minimum purity level of 70–80% of PET is required for an economically feasible process [57].

Natural polyamide fibres (i.e., wool and silk) are often used as a mono-material, and blends are limited. Wool can be combined with acrylic and still look like a woolen garment. Depending on the composition, it might be possible to also mechanically recycle blends of wool and acrylic, as is done with purely woolen garments. The manufactured polyamides, nylons, are frequently combined with other fibres such as cotton (e.g., in lace) or viscose (e.g., in knitted fabric). Similar approaches as for the combination of cotton and PET could be taken here.

5. Conclusions

Over the coming decades, a shift of the global textile value chain from a linear to a circular model is foreseen, driven by new regulations, the desire for resource efficiency, cost concerns and consumer demands. The textile industry is not prepared for this transition, as it lacks circular design competencies and efficient ways to recycle textile residues. For the textile industry to become more sustainable, knowledge of the origin and production of resources is important. Recycled feedstock, in addition to virgin renewable resources, is expected to form a significant part of the future resources to be used.

Textile fibre classifications are often made by origin and not by polymer type. Fibre properties are always described from the initial application and their behaviour during their lifetime through exposure, use or maintenance. Recyclability has not been a relevant variable until now.

For recycling purposes, we classified textile fibres based on their main chemical bonds, as fibres with the same kinds of bonds usually have similar chemical and physical characteristics. We distinguished cellulose, polyamide, polyester, polyurethane, polyolefin and polyacrylic as the principal polymer groups for textile fibres. In addition, a new classification of textile recycling technologies was designed based on the level of disassembly (fibre, polymer, monomer) combined with mechanical, physical and chemical recycling.

We showed that for fibre recycling by mechanical means, fibre length is the most crucial parameter. This type of recycling works best for cotton, linen, wool and acrylics. The quality of the recycled product is often lower than that of virgin resources and is highly dependent on the quality of the input stream. For polymer recycling by physical means, the molecular weight of the polymer and the ability to dissolve or melt are important. This type of recycling is best applied to cotton, linen, viscose, lyocell (dissolve) and nylon and PET (melt). The quality of the recycled product can approach virgin quality. For monomer recycling by chemical means, the ability to depolymerise the polymer to its monomeric building blocks is key, combined with an efficient recovery. This type of recycling is only suitable for polycondensation polymers such as nylon and PET. The most important advantage is that virgin quality can be achieved. Disadvantages include high costs and the needed large-scale production for economic feasibility.

For consumers, the choice between virgin and recycled products depends on the perceived benefits or disadvantages. Recycled content is often a synonym for low quality. However, here we showed that the final quality depends strongly on the type of fibre and technology used and that each fibre has its preferred recycling technology.

Textile recycling (especially post-consumer) is still in its infancy and will be a major challenge in the coming years. In general, there are good recycling options for mono-material residue streams, but the real challenge lies in blended textiles. The volume and composition of blended textiles allocated for recycling are often unknown, which was a limitation for this study. The plethora of fibre combinations added to this problem. Recycling blended textiles is possible to a limited extent with the currently available recycling techniques. However, when recycling is technically complicated, energy-consuming and expensive, it will most likely not become a profitable business, especially in combination with cheap virgin materials. For blended textiles, the perspective is promising for fibre blends within a single polymer group, while combinations of different polymers are undesirable.

To be able to Recycle, we must Rethink. Outdoor apparel brands have been experimenting with different design and material selection strategies that enhance the recyclability of their products. Examples are fabrics based on one type of fibre, fibres with better recyclability profiles, and creating a market demand for recycled materials [57]. Adoption of these approaches by other players in the textile industry is urgently needed, but initiatives are still scattered and small scale.

The intricate blending of different types of fibres that require different recycling strategies should be prevented, and the use of fibres that originate from renewable resources combined with good recycling options should be encouraged. Our methodology can help stakeholders in the textile industry to critically assess their production methods and the materials they apply. If the approaches we propose become more widespread, the recycling of fashion can grow, and the sustainability of the textile and fashion industry will improve.

Author Contributions: Conceptualisation, P.H. and H.B.; methodology, P.H. and H.B.; writing—original draft preparation, P.H.; writing—review and editing, H.B. and M.S. All authors have read and agreed to the published version of the manuscript.

Funding: This research was funded by Wageningen University and Research Knowledge Base Program *Towards a circular and climate neutral society* (KB34), project Recycling and end-of-life strategies for sustainability and climate (KB-34-011-001).

Conflicts of Interest: The authors declare no conflict of interest.

References

1. Carr, D.J.; Gotlieb, M.R.; Lee, N.-J.; Shah, D.V. Examining overconsumption, competitive consumption, and conscious consumption from 1994 to 2004. *Ann. Am. Acad. Politi-Soc. Sci.* **2012**, *644*, 220–233. [CrossRef]
2. Ellen MacArthur Foundation. A New Textiles Economy: Redesigning Fashion's Future. Available online: https://www.ellenmacarthurfoundation.org/publications/a-new-textiles-economy-redesigning-fashions-future (accessed on 5 February 2020).
3. Saidani, M.; Yannou, B.; Leroy, Y.; Cluzel, F.; Kim, H. How circular economy and industrial ecology concepts are intertwined? A bibliometric and text mining analysis. In Proceedings of the Symposium on Circular Economy and Sustainability, Alexandroupolis, Greece, 1–3 July 2020.
4. McDonough, W.; Braungart, M. *Cradle to Cradle: Remaking the Way We Make Things*; North Point Press: New York, NY, USA, 2010.
5. Rosa, P.; Sassanelli, C.; Terzi, S. Towards Circular Business Models: A systematic literature review on classification frameworks and archetypes. *J. Clean. Prod.* **2019**, *236*, 117696. [CrossRef]
6. Potting, J.; Hekkert, M.; Worrell, E.; Hanemaaijer, A. *Circular Economy: Measuring Innovation in The Product Chain*; PBL Publishers: The Hague, The Netherlands, 2017. Available online: https://dspace.library.uu.nl/handle/1874/358310 (accessed on 28 January 2021).
7. European Parliament. Directive 2008/98/EC of the European Parliament and of the Council of 19 November 2008 on waste and repealing certain Directives. *J. Eur. Union* **2008**, *312*, 312–330.
8. Piribauer, B.; Bartl, A. Textile recycling processes, state of the art and current developments: A mini review. *Waste Manag. Res.* **2019**, *37*, 112–119. [CrossRef] [PubMed]
9. Mather, R.R.; Wardman, R.H. *The Chemistry of Textile Fibres*; Royal Society of Chemistry: London, UK, 2015.
10. Sinclair, R. Understanding textile fibres and their properties: What is a textile fibre? In *Textiles and Fashion*; Sinclair, R., Ed.; Elsevier: London, UK, 2015; Chapter 1, pp. 3–27.
11. Wagner, M.; Heinzel, T. Human perceptions of recycled textiles and circular fashion: A systematic literature review. *Sustainability* **2020**, *12*, 10599. [CrossRef]
12. Sandin, G.; Peters, G.M. Environmental impact of textile reuse and recycling—A review. *J. Clean. Prod.* **2018**, *184*, 353–365. [CrossRef]
13. Tomaney, M. Sustainable textile production. In *Textiles and Fashion*; Sinclair, R., Ed.; Elsevier: Amsterdam, The Netherlands, 2015; Chapter 22, pp. 547–560.
14. Landi, S. *The Textile Conservator's Manual*; Routledge: London, UK, 1998.
15. Hiemenz, P.C. *Polymer Chemistry-The Basic Concepts*; Marcel Dekker: New York, NY, USA, 1984; pp. 211–215.
16. Lindström, K.; Sjöblom, T.; Persson, A.; Kadi, N. Improving Mechanical Textile Recycling by Lubricant Pre-Treatment to Mitigate Length Loss of Fibers. *Sustainability* **2020**, *12*, 8706. [CrossRef]
17. Gulich, B. Development of products made of reclaimed fibres. In *Recycling in Textiles*; Cambridge University Press: Cambridge, UK, 2006; Chapter 9.
18. Vadicherla, T.; Saravanan, D. Textiles and apparel development using recycled and reclaimed fibers. In *Roadmap to Sustainable Textiles and Clothing*; Springer: Berlin/Heidelberg, Germany, 2014; pp. 139–160.
19. Textile Exchange Preferred Fiber and Materials Market Report 2020. Textile Exchange 2020. Available online: https://store.textileexchange.org/product-category/corporate-fiber-materials-reports/ (accessed on 12 March 2021).
20. Frankenhuis. Textielrecycling Sluit de Keten in de Circulaire Economie voor Textielafval. Available online: https://www.frankenhuisbv.nl/ (accessed on 12 March 2021).
21. Wolkat Circulaire Textiel Recycling. Available online: https://www.wolkat.com/ (accessed on 12 March 2021).

22. Belda Lloréns Ecolife Products. Available online: https://www.ecolifebybelda.com/ecolife-products/ (accessed on 12 March 2021).
23. Geetanjali Woollens Dyed Wool, Cotton and Acrylic Blended Yarns. Available online: https://www.geetanjaliwoollens.com/yarns.html (accessed on 30 March 2021).
24. Ferre. From Waste to Yarn. A Circular Process. Available online: https://ferreyarns.com/products/ (accessed on 30 March 2021).
25. Velener Textil: Our No-Waste Model for Cotton Processing. Available online: https://www.velener.de/en/wecycled.html (accessed on 30 March 2021).
26. Lenzing. TENCEL™ x REFIBRA™ Technology. Available online: https://www.tencel.com/refibra (accessed on 30 March 2021).
27. Asahi Kasei Bemberg™—The Cupro Fiber from Asahi Kasei. Available online: https://www.asahi-kasei.co.jp/fibers/en/bemberg/ (accessed on 30 March 2021).
28. Renewcell Recycling Clothes Finally Works. Available online: https://www.renewcell.com/en/ (accessed on 30 March 2021).
29. Evrnu with NuCycl™ Technology. The World's Waste No Longer Has To Go To Waste. Available online: https://www.evrnu.com/nucycl (accessed on 30 March 2021).
30. Infinited Fibre. It's Time for the Textile Industry to Lose Its Virginity. Available online: https://infinitedfiber.com/ (accessed on 12 March 2021).
31. Ioncell. Research. Available online: https://ioncell.fi/research/ (accessed on 12 March 2021).
32. Saxion SaXCell Maakt van Gebruikt Textiel Nieuw Textiel in Pilotfabriek in Enschede. Available online: https://www.saxion.nl/nieuws/2020/april/saxcell-maakt-van-gebruikt-textiel-nieuw-textiel-in-pilotfabriek-in-enschede (accessed on 29 March 2021).
33. Bell, N.C.; Lee, P.; Riley, K.S.; Slater, S. Tackling Problematic Textile Waste Streams. *RESYNTEX*. Available online: http://www.resyntex.eu (accessed on 26 July 2018).
34. Shen, L.; Worrell, E.; Patel, M.K. Open-loop recycling: A LCA case study of PET bottle-to-fibre recycling. *Resour. Conserv. Recycl.* **2010**, *55*, 34–52. [CrossRef]
35. Datta, J.; Kopczyńska, P. From polymer waste to potential main industrial products: Actual state of recycling and recovering. *Crit. Rev. Environ. Sci. Technol.* **2016**, *46*, 905–946. [CrossRef]
36. VelenerTextil WETURNED®. PET-Bottles in Our Fabric, Not Our Oceans. Available online: https://www.velener.de/en/weturned.html (accessed on 29 March 2021).
37. Cumapol CuRe. Join the Polyester Rejuvenation Revolution! A Low Energy Recycling Solution for Used Polyester. Available online: https://www.cumapol.nl/curepolyester/ (accessed on 12 March 2021).
38. Ioniqa A Game Changer. Available online: https://ioniqa.com/ (accessed on 30 March 2021).
39. Jeplan Bring Technology. Making Clothing from Clothing. Available online: https://www.jeplan.co.jp/en/technology/fashion/ (accessed on 12 March 2021).
40. Teijin Ecopet. Quality from Waste. Available online: https://ecopet.info/en/ (accessed on 29 March 2021).
41. Eastman Polyester Renewal Technology. Available online: https://www.eastman.com/Company/Circular-Economy/Solutions/Pages/Polyester-Renewal.aspx (accessed on 30 March 2021).
42. Ambercycle. We Make Garments from Garbage. Available online: https://www.ambercycle.com/ (accessed on 30 March 2021).
43. Carbios Biorecycling. Available online: https://carbios.fr/en/technology/biorecycling/ (accessed on 30 March 2021).
44. Gr3n Long Life to Plastic Bottles. Available online: http://gr3n-recycling.com/ (accessed on 30 March 2021).
45. Russell, S.; Swan, P.; Trebowicz, M.; Ireland, A. *Review of Wool Recycling and Reuse*; Springer: Dordrecht, The Netherlands, 2016; pp. 415–428.
46. Mondragon, G.; Kortaberria, G.; Mendiburu, E.; González, N.; Arbelaiz, A.; Peña-Rodriguez, C. Thermomechanical recycling of polyamide 6 from fishing nets waste. *J. Appl. Polym. Sci.* **2019**, *137*, 48442. [CrossRef]
47. Mihut, C.; Captain, D.K.; Gadala-Maria, F.; Amiridis, M.D. Review: Recycling of nylon from carpet waste. *Polym. Eng. Sci.* **2001**, *41*, 1457–1470. [CrossRef]
48. DSM Cradle to Cradle, Nylon-6 en Caprolactam. Available online: https://www.duurzaammbo.nl/images/pdf/Presention%20DSM%20Nylon-6%20Recycling%20C2C%20Desso_WB.pdf (accessed on 26 March 2021).
49. Boer Group: Material Re-Use. Available online: http://boergroup-recyclingsolutions.com/textile-recycling/material-re-use/ (accessed on 12 March 2021).
50. Novetex.The Billie Upcycling. Available online: https://www.novetex.com/the-billie-upcycling/ (accessed on 12 March 2021).
51. Fulgar. Q-NOVA®. Available online: https://www.fulgar.com/eng/products/q-nova (accessed on 12 March 2021).
52. Aquafil. The ECONYL® yarn. Available online: https://www.aquafil.com/sustainability/econyl/ (accessed on 30 March 2021).
53. Van Dam, J.E.G.; Knoop, J.R.I.; Van Den Oever, M.J.A. Method for Removal of Polyurethane Fibres from a Fabric or Yarn Comprising Polyurethane Fibres and Cellulose-Based Fibres. WO Patent 2,020,130,825, 20 December 2020.
54. Ignatyev, I.A.; Thielemans, W.; Vander Beke, B. Recycling of polymers: A review. *ChemSusChem* **2014**, *7*, 1579–1593. [CrossRef] [PubMed]
55. Radhakrishnan, S. Denim recycling. In *Textiles and Clothing Sustainability*; Springer: Berlin/Heidelberg, Germany, 2017; pp. 79–125.
56. Deopura, B.L.; Padaki, N.V. Synthetic textile fibres: Polyamide, polyester and aramid fibres. In *Textiles and Fashion*; Sinclair, R., Ed.; Elsevier: Amsterdam, The Netherlands, 2015; Chapter 5, pp. 97–114.
57. Greenblue. Chemical Recycling; Making Fibre-to-Fibre Recycling A Reality For Polyester Textiles. 2018. Available online: http://greenblueorg.s3.amazonaws.com/smm/wp-content/uploads/2018/05/Chemical-Recycling-Making-Fiber-to-Fiber-Recycling-a-Reality-for-Polyester-Textiles-1.pdf (accessed on 26 March 2021).

58. Hopstaken, F.; van der Schalk, A.; van der Maesen, M.; Custers, F. Massabalans Textiel 2018. FFact: 27-3-2020. 2020. Available online: https://www.verenigingafvalbedrijven.nl/userfiles/files/rapport-massabalans-textiel-2018-2020.pdf (accessed on 30 March 2021).
59. van der Wal, E.; Verrips, A. Textiel als Secundaire Grondstof. Centraal Planbureau: 2019. Available online: https://www.cpb.nl/sites/default/files/omnidownload/CPB-Achtergronddocument-nov2019-Textiel-als-secundaire-grondstof.pdf (accessed on 26 March 2021).
60. Harmsen, P.F.H. (Wageningen Food and Biobased Research, Wageningen, The Newtherlands). Personal communication, April 2021.
61. Ishfaq, M. Infrared Technology and Its Applications Iin Textile Recycling Technology: Improving Sustainability in Clothing Industry. Master's Thesis, Lahti University of Applied Sciences, Lahti, Finland, November 2015.
62. H&M Foundation. Recycling Revolution—Inventing New Ways to Recycle Textiles. Available online: https://hmfoundation.com/project/recycling-revolution/ (accessed on 29 March 2021).
63. Worn Again: A World Where Resources Are Kept in Constant Circulation, Driving Economic, Social and Environmental Benefits. Available online: https://wornagain.co.uk/ (accessed on 29 March 2021).
64. Sustainordic. Blend Rewind. Available online: https://www.sustainordic.com/portfolio/items/blend-rewind/ (accessed on 29 March 2021).
65. Sodra. Once More. Available online: https://www.sodra.com/en/global/pulp/oncemorebysodra (accessed on 29 March 2021).
66. Textile World. Tyton BioSciences Developer of Recycling Technology For the Fashion Industry, Closes Series a Funding Round. Available online: https://www.textileworld.com/textile-world/knitting-apparel/2020/01/tyton-biosciences-llc-developer-of-recycling-technology-for-the-fashion-industry-closes-series-a-funding-round/ (accessed on 30 March 2021).
67. Block Texx. What We Do: S.O.F.T.—Separation of Fibre Technology. Available online: https://www.blocktexx.com/ (accessed on 30 March 2021).
68. Resyntex: A New Circular Economy Concept for Textiles and Chemicals. Available online: http://www.resyntex.eu/downloads (accessed on 30 March 2021).
69. Trash2Cash. Taking Waste & Making New Fibres. Available online: https://www.trash2cashproject.eu/ (accessed on 30 March 2021).
70. Palme, A. *Recycling of Cotton Textiles: Characterization, Pretreatment, and Purification*; Chalmers University of Technology: Göteborg, Sweden, 2016. Available online: https://publications.lib.chalmers.se/records/fulltext/246506/246506.pdf (accessed on 30 March 2021).

Article

A Delphi-Régnier Study Addressing the Challenges of Textile Recycling in Europe for the Fashion and Apparel Industry

Joséphine Riemens [1,2,3,*], Andrée-Anne Lemieux [2], Samir Lamouri [1] and Léonore Garnier [3]

1. LAMIH-UMR CNRS 8201, Arts et Métiers Sciences et Technologies, Institute of Technology, 75013 Paris, France; Samir.LAMOURI@ensam.eu
2. Institut Français de la Mode, Sustainability IFM-Kering Chair, 75013 Paris, France; aalemieux@ifmparis.fr
3. Fédération de la Haute Couture et de la Mode (FHCM), 75008 Paris, France; leonore.garnier@fhcm.paris
* Correspondence: josephine.riemens@ensam.eu or jriemens@ifm-paris.fr

Abstract: The increasing resource pressure and the expanding amount of textile waste have been rising recycling as a clear priority for the fashion and apparel industry. However, textile recycling remains limited and is therefore a targeted issue in the forthcoming EU policies. As the fashion industry is embedded in complex value chains, enhancing textile recycling entails a comprehensive understanding of the existing challenges. Yet, the literature review suggests only limited empirical studies in the sector, and a dedicated state-of-the-art is still lacking. Filling this gap, a Delphi study was conducted supplemented by the Regnier's Abacus technique. Through an iterative, anonymous, and controlled feedback process, the obstacles collected from the extant literature were collectively discussed with a representative panel of 28 experts, compared to the situation in Europe. After two rounds, the lack of eco-design practices, the absence of incentive policies, and the lack of available and accurate information on the product components emerged as the most consensual statements. Linking theory to practice, this paper aims to improve consistency in the understanding of the current state of textile recycling in Europe, while providing an encompassing outline of the current experts' opinion on the priority challenges for the sector.

Keywords: fashion; apparel; textile; recycling; challenges; circularity; sustainability

1. Introduction

With the rise of the globalization, the fashion industry has become a significant worldwide business (the global apparel market has been estimated in value to USD 1.5 trillion in 2020 [1]) based upon lengthy and geographically fragmented value chains [2,3]. Production has shifted to southern countries, mainly in Asia, while design, marketing, and distribution activities have remained located in advanced countries [4], with Europe representing an important apparel net importer (over EUR 80 billion of imported goods, mainly from China, Bangladesh, and Turkey, in 2019 [5]). Characterized by a constant growth and driven by the so-called "fast-fashion" phenomenon [3], the industry has resulted in tremendous environmental and social repercussions over the last decades [3,6]. Steadily scrutinized as a substantial impactful sector, a range of estimates have successively evaluated, depending on assumptions, the fashion industry to account for 4% [7], 8% [8], and up to 10% of global CO_2 emissions [3] with the main environmental impacts imputed to the raw materials and manufacturing activities [8,10,11]. Correlatively, this accelerated fashion consumption has led to the perception of clothes as disposable items, decreasing the garment's lifetime [9,12–16]. Such a completely linear or "take make waste" model has resulted in an expanding amount of discarded textiles in Europe [3,12,14,16]. While assessing used textile flows remains challenging [17], the amount of clothes bought in the EU has been estimated to have increased by 40% over the last decades [18,19]; Europeans consuming on average 26 kg and discarding 11 kg of textiles per person per year [20]. Considering the alarming climate state [21], decreasing the resource pressure is crucial to reduce the

environmental footprint of the sector while ensuring viable business models in the long term to face up to forthcoming scarcity [14,22]. Therefore, enhancing recycling has been standing as a critical priority in this wider transition towards a "circular economy" [9,14].

The concept of "circular economy" has raised interest since the late 1970s [14,23] and derives from connected background concepts such as the famed cradle-to-cradle design philosophy [23,24]. The most renowned definition has been framed by the Ellen MacArthur Foundation, where the economy is *"restorative and regenerative by design, and aims to keep products, components, and materials at their highest utility and value at all times"* [25]. Binding economic performance, social inclusiveness, and environmental resilience [26], this new industrial paradigm promotes material flows systems in which resources are optimized, thus enabling "sustainability" covering a broader framing [23]. Indeed, by promoting the reduction of resources extraction, the extension of the product lifecycle, and the minimization of waste production [27,28], the transition to circular business models has the potential to decrease the environmental impact of the industry [2,6,9,29,30]. However, the review suggests that companies are selectively implementing circular actions in their supply chains, rather than disrupting the entire business model [9] and addressing the main environmental impacts of their activities [12]. In fact, by bridging production and consumption activities [12,15,25], "circular economy" involves the implementation of several strategies usually suggested through the "R framework" and requiring radical systemic changes in how products and materials are manufactured, used, and disposed of [29–31]. A varying level of detail exist in the literature [32] and up to "10R principles"can be found namely refuse, rethink, reduce, reuse, repair, refurbish, remanufacture, repurpose, recycle, and recover [27,32,33]. Ordered by level of circularity, recycling rationally comes into play among the final options once the materials can no longer be reused, following the existing EU waste hierarchy of optimal treatment solutions [34].

However, it is estimated that only 12% of global material flows for clothing are recycled, mainly into "open-loop" applications [9] (such as insulation material, wiping cloths, or mattress stuffing), also referred to as "cascaded recycling" [35] or "downcycling" [36,37] because of their lesser economic value [38], while much higher rates are noticed in other industries [39]. Only less than 1% of clothing textile material would be recycled into new clothes, referred as "closed-loop recycling" or "textile-to-textile recycling", the 87% remainder ending up mainly in landfill or being incinerated [9]. Therefore, lifecycle-extending practices are not sufficient in themselves and must be enhanced, along with improving recycling practices [14]. Within this context, the sector is particularly targeted in the upcoming Green Deal strategy [40] aiming to achieve climate neutrality in Europe by 2050, as shown by the foreseeing EU Strategy for Textiles [41] ambitioning to boost the EU market for sustainable and circular textiles. The public consultations launched recently by the European Commission [41,42] reveal a wide range of dedicated policies to enhance textile recycling (i.e., separate collection mandatory by 2025, incorporation of recycled content, digital passport . . .).

However, despite increasing interest in the industry, textile recycling is still a limited research area [14], which has mostly been explored from a technical perspective, to improve textile recycling processes towards value-added products [36]. An increasing body of research has been also dedicated to the study of the environmental benefits of textile recycling [35]. However, while its improvement is consistently pointed out as requiring a "system-level change" with extensive stakeholders' collaboration [25], very few comprehensive studies have been conducted. The literature on the topic is fragmented, and empirical studies remain limited, which impedes an explicit evidence-based state-of-the art on the current challenges in the sector. Although a flourishing literature on textile-specific barriers to the implementation of the "circular economy" has advanced knowledge in the industry [12,29,32], these studies do not provide a systematic analysis of the existing issues related to the textile recycling value chain.

Subsequently, this paper aims to answer the following research questions: What are the current impediments in the textile recycling value chain? What are the priority challenges to address in order to enhance textile recycling in the sector? To achieve these research objectives, a systematic review was conducted to identify the recurring challenges acknowledged in the existing literature on textile recycling. Then, a purposely designed qualitative study using the Delphi method [43] and applying the Abaque de Régnier technique (Regnier's Abacus) [44] was implemented, referred to as "Delphi-Régnier" in the remainder of the paper. With an unprecedented panel of 28 experts in the field representing the different stakeholders of the recycling value chain, the reported challenges in the extant literature were translated into statements and iteratively discussed through this group communication technique, using a color grid to collect experts' opinions. The study is not strictly limited to textile-to-textile recycling but instead aims to explore the overall challenges related to the optimization of textile recycling, to better comprehend the current collection, sorting, and recycling system.

With regard to previous exhaustive research on the topic [14], the originality of the study resides both in the representativeness of the panel and in the Delphi-Régnier method used. This method differs from a simple interview and helps to advance empirical knowledge on textile recycling with the confrontation of the experts' opinions between the iterative rounds. Moreover, while barriers of textile recycling have started to be explored in the industry [12,14,29,32], this Delphi-Régnier study provides a holistic analysis of the contextual challenges and addresses the priority concerns in the sector, by allowing consensus and dissensus on the topic to arise. The research focuses on the European situation but still provides general findings advancing the literature on sustainability in the fashion and apparel industry. If the method has inherent limitations, the findings can support relevant initiatives for practitioners or policymakers and research opportunities to advance textile recycling in the sector.

The paper is structured as follows: Section 2 introduces the systematic literature review conducted. Section 3 describes the Delphi method and the Regnier's Abacus technique, together with the associated research steps. Section 4 is devoted to the presentation and discussion of the results. Section 5 introduces the limitations and research perspectives. At last, Section 6 concludes on the main findings.

2. Literature Review

A prior systematic review [45] was developed on the research problem, to explore the challenges of textile recycling with regards to the fashion and apparel sector. It was done using the databases Scopus, Science Direct, and Web of Science to guarantee an extensive coverage. As a relatively new and transdisciplinary area, we decided not to limit the scope of the review and include empirical studies, review papers, conference proceedings, and book chapters as long as they served the research objective. An additional search was done through Google Scholar to make sure that on-topic publications, unavailable on specific databases, were not omitted. Preliminary research suggested a limited academic literature, corroborated by Sandvik et al. [14]. Following the review protocol from preceding reviews on textile recycling [46], white papers and public reports from well-established organizations or institutions were also considered, after an extensive search on Google. By contrast, we noticed an extensive grey literature, especially several studies that have been published by practitioners for public institutions to improve resource management. However, the priority was given to peer-reviewed journals, and only the most relevant publications were cautiously included. Even if no limitation on the publication date was pre-established, research has significantly increased since 2005 with a clear rise in publications starting from 2018, and only the most recent papers were selected to ensure a contemporary state-of-the-art.

Keyword identification was challenging as in primary enquiries, engineering, material, environmental sciences, and chemistry fields prevailed, confirmed by the review performed by Shirvanimoghaddam et al. [36]. Initial extensive research was conducted using the combination (textile OR apparel OR garment) AND (recycling) AND (challenge OR obstacle OR limit OR complex OR barrier). Numerous papers on wastewater treatment were raised while our research focused on the narrowed understanding of textile recycling as the "*breakdown and reclaiming of textile raw materials to new ones*" [38]. Following an inductive and flexible process [45], the research protocol was progressively revised based on subsequent findings. We replaced ("recycling") with the keywords and Boolean connector ("circular" AND "waste") in the research string to broaden the scope and target papers on textile flows management that were relevant to the research problem. A total number of 1414 articles resulted from those two keyword equations, which were applied to titles, abstracts, and keywords for Scopus and Science Direct, and to "all fields category" for Web of Science.

The filtering process was another obstacle due to the fragmented literature. It was initiated by checking titles and abstracts, firstly excluding publications unrelated (1) to the aforementioned delineation of textile recycling and (2) to the textile and apparel industry. After removing the duplicates, this process resulted into 98 references. Then, full reading was performed to successively define inclusion and exclusion criteria. As a lot of papers were related to the aforementioned technical fields, we first only included articles providing overarching discussions on textile recycling, and we started classifying the reported challenges. Based on the initial findings, a second round of reading was carried out, and only publications either providing empirical knowledge or further supporting some of the challenges identified were incorporated, diminishing the sample to 20 publications. A flowchart of the literature review process is shown in Figure 1.

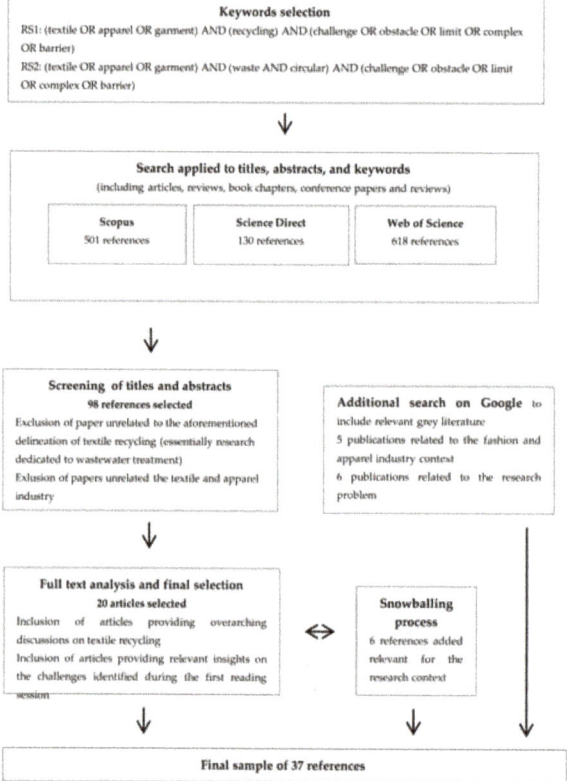

Figure 1. Flowchart of the systematic literature review.

The review starts with a general discussion on the key findings, supported by a mapping eliciting the outlined challenges and their recurrence rate in the academic literature, presented subsequently in Table 1.

Table 1. Reported challenges with occurrences in the academic literature.

Title	Document Type	Source	Year	Collection Infrastructures	Sorting Process	Information on Material and Chemical Content	Ease of Disassembly	Innovative Recycling Technologies	Profusion of Material Blends	Yarn and Fabric Construction	New Materials Introduction	Use of Chemical Substances	Use Conditions	Recycled Output Quality	Output Recyclability	Connection between Supply and Demand	High Costs Associated to Recycling	Research and Investment Costs	Market Opportunities	Awareness and Education	Collaboration between Stakeholders	Guidelines, Regulations and Supporting Policies	Environmental Performance Assessment
Towards a Conceptual Framework of Sustainable Practices of Post-consumer Textile Waste at Garment End of Lifecycle: a Systematic Literature Review Approach [47]	Review	Sustainability	2021		x				x											x			
Textile Recognition and Sorting for Recycling at an Automated Line Using Near-Infrared Spectroscopy [48]	Article	Recycling	2021	x	x																		
Analysis of the Polyester Clothing Value Chain to Identify Key Intervention Points for Sustainability [49]	Review	Environmental Sciences Europe	2021		x			x	x													x	
Closing the Loop on Take, Make, Waste: Investigating Circular Economy Practices in the Swedish Fashion Industry [12]	Article	Journal of Cleaner Production	2021					x	x													x	
Apply DEMATEL to Analyzing Key Barriers to Implementing the Circular Economy: An Application for the Textile Sector [29]	Article	Applied Sciences	2021			x		x						x			x			x		x	
Circular Economy of Post-consumer Textile waste: Classification Through Infrared spectroscopy [50]	Article	Journal of Cleaner Production	2020	x	x												x						
Death by waste: Fashion and Textile Circular Economy Case [30]	Review	Science of the Total Environment	2020						x			x	x	x	x							x	
Close the loop: Evidence on the Implementation of The Circular Economy from the Italian Fashion Industry [28]	Article	Business Strategy and the Environment	2020					x	x									x			x		
A Conceptual Framework for Barriers of Circular Supply Chains for Sustainability in the Textile Industry [32]	Article	Sustainable Development	2020	x	x	x	x	x	x			x		x		x	x				x	x	
The Circular Economy in the Textile and Apparel Industry: A Systematic Literature Review [51]	Review	Journal of Cleaner Production	2020			x	x	x	x			x	x		x	x	x					x	
Circular Fashion Supply Chain Through Textile-to-textile Recycling [14]	Article	Journal of Fashion Marketing and Management	2019	x	x	x	x	x	x	x	x	x	x	x	x	x	x		x	x	x	x	
A Review of the Socio-economic Advantages of Textile Recycling [7]	Review	Journal of Fashion Marketing and Management	2019	x		x	x	x					x			x	x		x	x		x	
Improving Recycling of Textiles Based on Lessons from Policies for Other Recyclable Materials: A Minireview [39]	Review	Journal of Cleaner Production	2019	x	x	x	x	x	x			x				x	x		x	x	x	x	
Textile Recycling Processes, State of the Art, and Current Developments: A Mini Review [52]	Review	Waste Management & Research	2019		x		x	x	x					x									
Environmental Prospects for Mixed Textile Recycling in Sweden [37]	Article	ACS Sustainable Chemistry & Engineering	2019		x		x	x				x	x	x									x
Effects of Cotton Textile Waste Properties on Recycled Fiber Quality [53]	Article	Journal of Cleaner Production	2019							x		x		x						x			
Evaluation of a European Textile Sorting Center: Material Flow Analysis and Lifecycle Inventory [54]	Article	Resources, Conservation and Recycling	2019					x	x										x		x	x	
Circular Economy—Challenges for the Textile and Clothing Industry [6]	Review	AUTEX Research Journal	2018	x				x	x			x							x	x	x	x	x

Table 1. *Cont.*

Title	Document Type	Source	Year	Collection Infrastructures	Sorting Process	Information on Material and Chemical Content	Ease of Disassembly	Innovative Recycling Technologies	Profusion of Material Blends	Yarn and Fabric Construction	New Materials Introduction	Use of Chemical Substances	Use Conditions	Recycled Output Quality	Output Recyclability	Connection between Supply and Demand	High Costs Associated to Recycling	Research and Investment Costs	Market Opportunities	Awareness and Education	Collaboration between Stakeholders	Guidelines, Regulations and Supporting Policies	Environmental Performance Assessment
Developing a National Programme for Textiles and Clothing Recovery [55]	Article	Waste Management and Research	2018		X		X	X	X					X	X		X		X				
Open- and Closed-loop Recycling of Textile and Apparel Products [38]	Book chapter	Handbook of Life Cycle Assessment (LCA) of Textiles and Clothing	2015	X	X			X	X					X	X					X	X	X	X

2.1. Complexity of the Textile Recycling Value Chain

Consistent with the definition from the EU Waste Framework Directive [56], textile recycling is essentially designated from a technical perspective as the *"breakdown and reclaiming of textile raw materials to new ones"* [25,35,36] involving a degree of deconstruction [49]. However, textile recycling is actually embedded in a broader and complex textile flow management system involving several preceding stages.

Starting with the collection, textiles are usually classified depending on their collection point along the value chain: textile flows are either designated upon "pre-consumer" if collected at the industrial level or "post-consumer" if collected after use [25,32,57]. "Pre-consumer" usually comprises all types of waste from companies resulting from manufacturing and distribution activities, from textile offcuts to unsold products, whereas "post-consumer" refers to used articles discarded by consumers once no longer wanted. Upstream, some also distinguish between the "post-industrial" (side-effect of clothing manufacture) and the "pre-consumer" (inferior quality garments or unsold product at the retail stage) feedstocks depending on the source in the value chain [16,58]. Nevertheless, while the "pre-consumer" fraction in general is underreported [59], the post-consumer part entails the major challenges for the sector. Overall, 15%–20% of clothes have been estimated to be collected for reuse and recycling practices within Europe [35,46], and important disparities of collection schemes are addressed [9,46]. Attempting mappings on textile flows highlights collection rates varying from 22% (Sweden) [60], 39% (France) [61], and 45% (Denmark) [60], up to 75% in Germany [9] of overall textiles put on the market.

Once collected, items undergo thorough manual sorting, which is resting upon a business model towards the second-hand market [46,54] and the quest for high-quality clothing [46], also referred as "diamonds" [57] or "cream" [17,61]. While oversimplified, the reusable feedstock is separated from the nonreusable stock. A significant fraction is exported abroad, while the most valuable part is directed to fine sorting for nearby markets [11,40,41,54]. Still, for the majority being sold in global markets, the question of their subsequent treatment is left untackled, which explains the current low estimates of textile recycling [9]. Moreover, some exports markets are becoming saturated [54,62], and the quality of reusable textiles appears to have fallen [54,61], a situation further exacerbating the primary issue of textile recycling for the coming years in the sector.

Following reuse, the nonreusable fraction is directed across different recovery avenues (recycling, energy recovery, or incineration) or landfilled with varying rates depending on the countries. In preparation for recycling, textiles are further sorted manually based on their composition [61] and color [54], vis-à-vis recyclers' specifications requirements. In addition, external components (i.e., metallic parts, buttons, zippers) are disassembled prior to recycling mostly through a manual process [38,46]. To put this into perspective, in France, where an Extended Producer Responsibility (ERP) has been in place since 2007 for clothing, linen, and shoes [55], 56.5% of the items sorted in facilities under contract are reused, mainly overseas (95%), 33.3% is recycled, 9.1% is recovered for solid fuel, 0.7% is incinerated for energy recovery, and only 0.4% is disposed, either through incineration or landfill [61]. The ERP is a policy tool for encouraging and enabling recycling by giving financial and/or physical responsibility to producers for treating and disposing of post-consumer products [17], which is particularly considered in the upcoming EU strategy for the textile sector [41].

Even when nonreusable textiles are recycled, the main destinations remain "open-loop" applications such as wipers, nonwoven, or insulation felts towards other industries [9,14,63]. Therefore, a technological barrier is commonly outlined, and a strong technical emphasis has been featured in the literature on textile recycling.

2.2. The Technological Barriers of Textile Recycling

The lack of technologies to support the development of textile-to-textile recycling is identified as a crucial persisting barrier, as attested by the recurrence rate documented in Table 1. Indeed, textile recycling is predominantly based on established mechanical recycling [9,26]. As regards to the existing processing routes, a typical classification is made between mechanical and chemical recycling [38,58].

Mechanical recycling implies cutting the fabric into smaller pieces and then progressively shredding them through a rotating drum until reaching a fibrous state, suitable for re-spinning into new yarn or for other manufacturing products. Less common, "thermal recycling" is occasionally specified [36] to recycle synthetic fibers by melt extrusion, regularly mentioned under "mechanical recycling" as implying prior mechanical processing [35]. The process can be used to recycle fabrics made from natural fibers as well as synthetic fibers [38]. However, the shredding of the fabric during the mechanical process shortens the fiber length and thus, reduces the quality output [37,51,53]. In "fiber recycling" [9], recycled fibers need to be blended with virgin fibers to reach a satisfactory quality level for yarn spinning [38,46]. Therefore, pre-consumer flows are considered more suitable than post-consumer waste for textile-to-textile recycling, as they are more consistent in quality and color [35,38], with exceptions such as wool, made up of long fibers [58]. Polyester and other thermoplastics can be recycled through thermal recycling, also referred as "mechanical polymer recycling" [9], during which the garments are cut and granulated into PET pellets by applying heat [49] with a certain loss of quality [49,64]. Despite being technologically feasible, this process is however not yet applied at scale [9]. Moreover, apparel textiles are often composed of fiber blends, thereby introducing the challenge of separation [14,58,65] and limiting the range of recycled commodities [51]. Consequently, the main destinations remain shredding the textiles to fiber for applications such as nonwoven or insulations, explaining the negligible "textile-to-textile" recycling share previously mentioned [38]. As such, the distinction between "open-loop" and "closed-loop" aims to reflect the outcome level of refinement [35]: "closed-loop recycling" designates "material recycling for a more or less identical product" [17,38] while "open-loop recycling" appoints processes in which the material is used in another product [17,38], typically towards other industries. In this perspective, a body of scientific research has explored potential applications via mechanical recycling, including composites, sound absorber, or thermal insulation [36].

By contrast, "chemical recycling" implies a higher degree of processing through depolymerization (process of breaking polymeric bonds, for synthetics fibers only) or dissolution (for natural or synthetic cellulosic fibers) [35]. However, chemical recycling is still limited [9,62]. Several innovative processes have been developed [52], especially towards pure cotton and cotton–polyester blends [9]. However, the current technological state is ambiguous, the review suggesting a lack of technological maturity, [9] a deficient economic viability, [9] or the need for investment in research and development [46]. Consequently, the market share of recycled fibers was estimated in 2020 at 8.1% of global fiber production, but such incorporation of recycled fibers mainly derives from the recycling of PEF bottles [66] established since the 1970s [38,67]. In addition, it must be noted that this typology appears oversimplified, as recycling operations often comprise a mix of mechanical, chemical, and thermal processes [35]. Therefore, further classifications have been proposed according to the level of disassembly [9,35] and more recently to the polymer structure of the fibers [27] to promote a clearer communication on the technological progress of textile recycling.

Another technological barrier resulting from the sorting stage is reported in the literature [9,17,47,48,50,68]. As previously outlined, sorting is a manual and costly phase, and therefore, there is a need for an economically viable and effective way to recognize and sort textile materials, to further advance textile-to-textile recycling as requiring homogenous feedstocks [48,50]. This current manual sorting is lacking reliability to recognize and sort the items according to their material content, jeopardizing the subsequent recycling process [48]. The item's composition is often varying over its lifetime, and the care labels are often cut-off or inaccurate. According to a study conducted by Circle Economy, up to 41% of labels on blended materials contain incorrect information [68].

2.3. Systemic Challenges to Enhance Textile Recycling

As part of this complex textile flow management system, textile recycling involves a myriad of other stakeholders [9,35,47,57] such as charities, municipal waste-collection services, or resell organizations. Only a few mappings of textile flows can be found in the academic literature [16,57], and empirical studies investigating textile flows in these sorting facilities are lacking [54]. However, changes are required throughout the whole value chain, as the performance of this eco-system determines the amount of collected textiles than can be further reused and recycled [54]. Therefore, beyond technical matters [14], numerous obstacles are encountered in the optimization of textile recycling, entailing it as a systemic challenge [14,38].

Several studies, both theoretical [16,51] and empirical [26,29,32], have started to explore the textile-specific barriers related to the implementation of circular economy. While some studies provide in-depth analysis [29,32], they investigate the impediments related to the circular economy as a whole and therefore do not enable a holistic and systematic analysis of the recycling value chain. To the best of our knowledge, the most recent extensive research was carried out by Sandvik et al. in 2019 [14]. The study took a qualitative approach through semi structured interviews with 11 stakeholders and investigated the drivers, inhibitors, and enablers of creating a textile-to-textile recycling system in the Scandinavian fashion industry. The main inhibitors found were the limited recycling technology to separate materials, high costs of research and development, supporting logistics, and complexity of supply chains including multiple stakeholders. However, while extending knowledge on the types of systemic and technological changes required, the study does not allow to leverage the amplitude of opinions and the sample's lack of representativity. Moreover the study only focuses on textile-to-textile recycling, while both closed-loop and open-loop applications have their importance to reintegrate the increasing nonreusable feedstock [38].

The remainder of the research remains theoretical and only looks over some challenges, without further investigation. Therefore, the literature is scattered, as shown by the fluctuated attendance rate in Table 1. Although some challenges stand out more than others, the current analysis is unable to provide a clear vision on the research problem. Among the most comprehensive articles, Filho et al. [46] investigated the socio-economic advantages of textile recycling and provided the most systematic review we found on textile recycling. The review explored the socio-economic advantages of textile recycling and pinpointed barriers to the optimization of textile recycling without specifying if they were arranged by importance: (1) economic viability, (2) composition of textile products, (3) non availability of recyclable textile materials, (4) technological limitations, (5) lack of information and limited public participation, and (6) poor coordination, and weak policies and standards. In comparison, Holes and Holes investigated effective policies and incentives worldwide for increased recycling from other sectors to reveal opportunities for extrapolation to the textile sector. A brief overview over the challenges identified is exposed, respectively: (1) the lack of incentives to motivate consumers for recycling; (2) informational and educational programs; (3) collection options of textile waste; (4) absence of policies and regulations concerning textile recycling; (5) variety of textile fibers and chemicals.

Conversely, several reports and white papers on textile recycling [9,17,62] have been published over the years, providing flourishing grey literature on the topic. Especially, the Ellen MacArthur Foundation identified four priorities to improve textile-to-textile recycling [9]: align clothing design and recycling processes, pursue technological innovation to improve the economics and quality of recycling, stimulate demand for recycled materials, and implement clothing collection at scale. Several other publications have addressed the challenges of textile recycling. For instance, Elander and Ljungkvist investigated through in-depth interviews with fashion companies, textile sorters, and recyclers 43 critical aspects for increasing textile-to-textile recycling [63]. Roos et al. [17] compiled a state-of-the-art on existing technologies while outlining the important factors for the future of textile recycling.

Based on the reported challenges in the academic literature, a thorough mapping was framed, as shown in Table 1. The mapping essentially emphasizes the academic literature to outline the studies conducted so far while highlighting the fragmented state of the research. This literature review supported the implementation of our Delphi-Régnier study, and the challenges are further elaborated together with the findings in the dedicated section.

3. Research Method

Considering the purpose of the research, the combination of the Delphi method and the Regnier's Abacus technique emerged as the most relevant approach. By allowing consensus and dissensus on a topic to arise, this research approach helps to advance empirical knowledge on textile recycling while outlining the priority challenges to address in the sector with regards to previous research [14,39,46]. It also appeared as the most viable approach to evaluate a series of wide-range assumptions on the topic with a wide panel of experts and generate valid results in a timely manner [43]. The appropriateness of the method was also evaluated within the surrounding industry context [43]. As a spotlighted issue within the sector, anonymity was essential to ensure participation and minimize the risk of bias. In addition, confronting opinions on the topic was decisive to achieve the research objectives, but direct communication between the different experts could have blocked open reflection, due to distinct business interests and unaccustomed exchange of views of the different stakeholders involved.

Hereinafter, the Delphi method and the Regnier's Abacus technique are introduced, followed by a dedicated focus on the sequential research process of the study.

3.1. Delphi Method and Regnier's Abacus Technique

3.1.1. Delphi Method

The so-called "Delphi" is a well-acknowledged and widely used method for consensus-building on specific topics, solicited from experts through an iterative multistage process [43]. Named in reference to the Greek oracle, this method was pioneered by the Air-Force-sponsored Rand Corporation in the 1950s and later developed by Dalkey and Helmer [69] to forecast the potential of military technology [70]. At the time, quantitative simulation was primitive and conventional face-to-face experts' consultations were unable to provide reliable results [70]. Structured around an anonymous and controlled feedback process, this method was designed to minimize the typical group interaction shortcomings (e.g., influences of dominant individuals, group pressure for conformity, standing within a profession) and ensure independent expert's judgments [43].

While inherently flexible, the Delphi method consists of the following sequential procedure [43]:

- A questionnaire is submitted anonymously to the panel of experts (also referred as "participants", "respondents", or "panelists").
- Responses are counted and processed (also referred as "round" or "iteration" with the submission of the questionnaire).
- Based on these responses, a refined questionnaire is elaborated and submitted to the same participants, along with a summarized report of the prior iteration. Through this controlled feedback process, experts can reassess their initial judgments and additional insights can be provided, allowing the information collected to be thoroughly clarified as the rounds progress.
- This process is iterated until common tends are achieved and are precise (consensus and dissensus).

By facilitating exchanges of opinions, the Delphi process has been progressively used beyond forecasting in diverse application areas such as program planning, needs assessment, policy determination, and resource utilization [43].

With respect to the fashion industry, the methodology has recently received interest in research on sustainability challenges [71] and the underlying issue of traceability [72,73].

Especially, the method appears suitable to investigate complex and multidisciplinary problems [72] by allowing to correlate informed judgements on a topic spanning a wide range of disciplines, to explore or expose underlying assumptions or information leading to differing judgements, and to seek out information that may generate a consensus [74].

The Delphi method being conditioned by its use; different analysis techniques can be used to interpret the data [43] and thus, various forms of Delphi have been applied in the literature. As part of our study, the information investigated is essentially qualitative. Therefore, in coherence with our research objectives, we decided to use the Regnier's Abacus formalized technique, particularly suitable for collecting and processing qualitative information [75].

3.1.2. The Regnier's Abacus Technique

The Regnier's Abacus was developed in the 1970s by P. François Régnier in the medical field, to offset the usual drawbacks of working groups. By using a color panel to collect and share opinions on a topic, this visual communication technique was designed to promote constructive debates and facilitate decision-making. Initially assembled with colored cubes representing the decision scale [44], the available digital tools have further enhanced the possibilities associated with this method by allowing to collect and process data instantly. Thus, the abacus continues to attract interest and is used for many applications, ranging from forecasting research to simple projects evaluation [76].

The modalities are very simple as they are based on the "traffic signals" logic. Statements, also referred to as "items", are previously defined on the topic investigated to provide an opening framework for reflection, though in precise, concrete, and relevant terms [75]. Experts are invited to react on each item, by selecting one of the following seven colors reflecting the hierarchy of possible opinions:

- Green: the expert strongly agrees with the statement;
- Light green: the expert agrees with the statement;
- Orange: the opinion of the expert is mixed;
- Light red: the expert disagrees with the statement;
- Red: the expert strongly disagrees with the statement;
- White: the expert cannot answer;
- Black: the expert does not want to answer.

The light color indicates a transparency in answers, while white and black reflect opacity [75]. Therefore, in contrast to other data-gathering or analysis techniques, the colorful matrix produced enables a comprehensive and immediate perception of opinion. To complete this study, the open-source solution Color Insight (http://colorinsight.fr/, accessed on 5 February 2021) was used to generate the questionnaires, collect answers, and process information. An overview of the Color Insight voting platform is provided in the Figure A1 (Appendix A).

The respective steps involved in conducting the study are described in the following subsection.

3.2. The Implementation of the Delphi-Régnier Study

Considering proper planning management and motivation are key when conducting a Delphi study [43], particular attention has been paid to the distinctive research steps, due to the wide variety of stakeholders involved in the panel.

To give an insight of the study's implementation, Table 2 illustrates the different steps involved in the research process, while the formulation of the initial statements, the selection of experts, and the rounds procedure are further explained successively.

Table 2. Steps involved in the implementation of the Delphi-Régnier study.

Step 1	Literature review on textile recycling challenges from a fashion and apparel perspective Identification and codification of the challenges reported in the literature Formulation of the initial statements
Step 2	Identification and selection of the experts: to ensure a representative panel of the whole value chain, the panel was formed on the following stakeholder groups: • G1: Fashion and apparel companies • G2: Textile manufacturing companies • G3: Collecting and sorting companies • G4: Recycling companies • G5: Professional associations and organizations • G6: Research and innovation institutions and academia
Step 3	Submission of the first questionnaire to the experts panel for consultation
Step 4	Results review and development of the refined and additional statements
Step 5	Second round of statements presented to the experts panel for consultation
Step 6	Final comments review

3.2.1. Phase 1: Building the Opening Questionnaire

The first key step was building the initial questionnaire submitted to the experts in the first round, as it significantly guides the areas in which the Delphi-Régnier study will generate ideas [75]. We decided to establish the questionnaire beforehand, in order to target the most competent experts, based on the pre-established statements. During the literature review, a coding process was performed identifying and grouping by coherent themes the challenges reported. Based on this matrix, presented in Table 1, a set of 23 statements, also referred to as "items" in this paper, was formulated, as shown in Appendix B. All statements were carefully examined iteratively by the authors to reduce the risk of misinterpretation.

3.2.2. Phase 2: The Constitution of the Experts' Panel

The constitution of the panel was the second key step. While no standard number of experts is defined in the literature, it is considered that a Delphi study can generate results from 11 experts [75,77]. Moreover, it likely depends on the research purpose and the reference groups involved [43]. If the sample size is too small, the pooling of judgments will not be representative enough, while if the sample is too large, the completion of the study may be infringed by potential low response rates, due to the inherent time-consuming feedback process [43]. Within our research context, this step was particularly challenging because of the fragmented nature of the textile and fashion value chain, all the way to the recycling stage. Although recruiting experts from homogeneous background entails the risk of producing biased results [43], the involvement of the different stakeholder groups was essential to confront their views on this system-level matter. Given this unprecedent representativeness, we assumed that consensus observed would rather provide substantial results for the sector.

To select the most appropriate individuals, authors recommend proceeding through a rigorous nomination process without further specifications [78]. Therefore, we initially defined relevant stakeholder groups reflecting the different actors involved in the recycling value chain. We also considered it relevant to include academia, confederations, and other active institutions specialized in the field of textile recycling. Based on these stakeholder groups, we progressively built a database of experts by soliciting professional recommendations and by conducting a review process, using a number of sources, including media articles, available reports, web searches, or industry association membership. We anticipated that motivation would be key to the successful implementation of the study [43] and thus, we decided to initially send an email to each nominated expert with a summary document explaining the research objective, the method, and the timeframe, to gauge interest and availability to participate in the study. As we received responses or not from the experts, we continued this process reiteratively until reaching a sufficiently representative panel that would allow the convergence of ideas on the topic.

In total, we contacted 39 experts and we obtained the confirmation of 28 experts, as shown in Table 3. To ensure the most competent experts within the specialized area of knowledge and ensure the quality of the results [43], we mainly selected senior skill-level profiles. The panel is evenly split between women (61%) and men (39%), as well as between the stakeholder groups:

- Group 1: Fashion and apparel companies (18%);
- Group 2: Textile manufacturing companies (11%);
- Group 3: Collecting, sorting, and trading organizations and companies (25%);
- Group 4: Recycling companies (14%);
- Group 5: Institutions and policymakers (14%);
- Group 6: Research, support, and innovation organizations (18%).

Table 3. Panel of the 28 experts established following the selection process.

Stakeholder Group	Type of Stakeholder	Expert's Position
G1 Fashion and apparel companies	High-end segment	Head of Circularity
	High end segment	Head of Recycling Activities
	Mid-range segment	Head Quality and R&D
	Sportswear	Textile Recycling leader
	Workwear	Owner
G2 Textile and spinning manufacturing companies	Spinning mill (mechanical recycling wool, cotton/polyester, polyamide)	Sales Manager
	Textile company (silk)	Textile Designer Head of R&D
	Textile company (wool)	Head of Innovation and Sustainability
G3 Collecting, sorting, and trading organizations and companies	Charity organization	Head of Textile Mission
	Social integration company	Textile Recycling Project Manager
	Industrial group (comprising recycling activities)	Head of Collection Program
	Solution for brands for reuse, recovery, and recycling	Chief Executive Officer (CEO)
	Solution for brands for reuse, recovery, and recycling	Chief Executive Officer (CEO)
	Solution for brands for reuse, recovery, and recycling	Chief Executive Officer (CEO)
	Trading and tracking platform for industrial textile waste	Sales and Marketing Lead
G4 Recycling companies	Chemical recycling company	Process Technician
	Chemical recycling company	Chief Marketing Officer & Head of Investor Relations
	Mechanical recycling company	Chief Executive Officer (CEO)
	Mechanical recycling company	Chief Sustainability & Marketing Officers
G5 Institutions and policy officers	Professional Confederation	Policy Officer Sustainable Business
	Professional Confederation	Senior Policy Officer
	Professional Confederation	Recycling Project Manager
	Organization managing end-of-life products	Head of Innovation and Recycling
G6 Research, support, and innovation organizations	Non-Profit Organization promoting Circular Economy	Senior Policy Officer
	Non-Profit Organization promoting Circular Economy	Research Analyst Circular Textiles
	University	Program Manager for Sustainable Textiles
	Innovation Research Centre	Business Manager Circular Economy
	Industrial Cluster	Business Manager

3.2.3. Phase 3: The Round's Procedure

The initial questionnaire of 23 statements was sent to the experts through a dedicated voting link via Color insight, on which they were required to provide their opinion based on the Régnier Abacus color grid, as shown in Figure A1 (Appendix A). To avoid any interpretation during the analysis, participants were also asked to justify each of their answers with a short comment. Moreover, at the end of the questionnaire, experts has the opportunity to submit additional statements if some were missing, thereby strongly enriching the study. As suggested in the literature [43], the timeframe was a challenging aspect to maintain the subject motivation during the completion of the study, notably with the high number of participants. To prevent non-response and avoid a discontinuity in the answers between the rounds, experts were given two weeks to reply, and the same period was planned for the questionnaire review.

Resulting from the color votes proposed by the experts on each statement submitted in this first questionnaire, an item matrix was generated, as shown in Figure A2 (Appendix D). The items matrix enables to overview, from top to bottom, the most favorable statements to the most unfavorable, with dissensus in the middle of the picture. The matrix was generated according to the "classic mode" proposed by Color Insight: the color weights are 5 for dark green, 4 for light green, 3 for orange, 2 for light red, and 1 for dark red.

Throughout a review process, the authors examined and synthesized the votes and comments from this first round. As there is no standard threshold in the literature, it was decided that a consensus was reached once 60% of answers in favor (green) or against (red) was observed. Based on the comments, 3 statements were added and removed respectively, and all the other items were carefully reworded accordingly. Such modifications were clearly indicated in the second questionnaire, by the respective mentions of "new item" and "edited item" in front of each statement.

The second round was launched with the same participants and included the following:
- The refined questionnaire of 21 items provided in Appendix C.
- A summary document presenting the first-round results, to give the occasion to the experts to confront and reassess their previous answers.
- A full report with the breakdown of votes and the anonymized comments for each item was added to minimize the distortion risk, commonly described in the literature [43].

No new comments emerged during the second round, and the major opinion trends among the experts were confirmed. Therefore, we considered that the majority of convergence was reached, and we decided to end the study.

Only positive consensus was reached following both questionnaires, the remaining statements being subject to dissensus among experts. As shown by the respective matrix Figure A2 (Appendix D) and Figure A3 (Appendix E), the results are not highly pronounced, which complicated the analysis. Therefore, the examination of votes and comments was performed through iterative reviews by the steering committee to limit as much as possible the subjective bias in the interpretation. Nevertheless, favorable consensus and disagreement stand out to a greater extent in the second round. As a result, 10 statements reached positive consensus, out of the 21 submitted, despite the heterogeneity of the panel, and the supporting comments provided further insights, described in the findings.

It should be noticed that 23 of the 28 experts from the first round responded to the revised questionnaire in the second round. While this drop in participation is one of the shortcomings of the research method [43], the representativeness of the panel was maintained, and thus, the findings remained coherent with the preceding results.

4. Findings and Discussion

Following the Delphi-Régnier study, the steering committee synthetized the answers provided by the experts. Considering the refinement and addition of statements, based on received comments in the second round, we attempted to group the opinions by coherent themes. Derived from the study conducted by Elander et al. [63] in 2016, the following five categories were identified as the most relevant for aggregating the statements while ensuring a sufficient level of detail. Although closely connected, the categories encompass critical aspects of textile recycling, thereby allowing to evaluate the reviewed challenges in relation to the value chain described in the first section of the paper. In addition, a significant "organizational" issue emerged during the completion of the study, which resulted in the inclusion of a fifth distinctive category entitled "coordination".

- Product and material input
- Information
- Technologies
- Markets
- Coordination

The key findings are discussed accordingly, using the subsequent formula to reference the items: (RY-IX)—Y referring to the round's number and X to the statement's position in the items' matrix provided in Appendix D and the Appendix E.

4.1. Product and Material Input

4.1.1. The Lack of Eco-Design Practices

The lack of eco-design practices in the sector is highlighted as a major challenge (R2-I1). While addressed through several design features in the first questionnaire, it was distinctively mentioned in a number of comments, and it emerged as the highest positive consensus of the study (82%). Fashion companies currently do not integrate recycling in the design process, while the end-of-life of the product should be strategically integrated in the design process [12,15,51]. Moreover, eco-design must be addressed as a thought-confronting process including all the value chain, considering both durability and recyclability, which can be conflicting in practice. Several authors highlighted such an issue [14,63], and Sandvik et al. advanced the notion of "conditional design" [14] requiring further understanding of the use context of clothes.

Among the most impactful design features, the profusion of material blends reached the strongest favorable consensus (75%), consistent with the findings from the literature review, while it was slightly mitigated in the second round (R1-I1 and R2-I8). Textile materials can mostly be recycled with current mechanical recycling technologies, but material blends restrain recycling applications mainly to very low value applications. Yet not all blends are problematic, and some are necessary to provide long-lasting products, raising the potential contradiction between durability and recyclability. Moreover, it is likely to be overcome with adapted and efficient technologies, partly explaining the nuanced opinions observed.

The presence of external accessories, also referred to in practice as "hard points" (such as buttons, zippers, rivers ...), emerged as an important disruptor of textile recycling (75%), particularly in the first round (R1-I4). Certain external elements (i.e., care labels) are problematic in certain recycling pathways, and especially for closed-loop recycling by contaminating the process, depending on their composition. This is equivalent to the preceding statement relating to material blends. In addition, the presence of external accessories is causing cost inefficiencies due to the manual dismantling process. Yet, other obstacles seem more critical, and the statement received less favorable votes (57%) in the second round (R2-I14).

The other design features enquired about, which are less prevalent in the literature, do not stand out. Yarn and fabric construction can influence the appropriate recycling route, especially for closed-loop applications, but to a lesser extent than material blends (R1-I19). The introduction of new materials is not considered as a potential inhibiting factor (R1-I21). There is rarely an introduction of new materials, and it likely depends on which materials. Even new materials form part of a wider fiber family and can be grouped with similar fiber types, as highlighted by one expert.

Finally, the initial fiber quality is more decisive than the use conditions (i.e., wearing and washing), especially to enable spinnability in closed-loop applications through mechanical recycling, but not with chemical recycling (R1-I13).

4.1.2. The Presence of Chemical Substances

Almost reaching positive agreement (57%) in the first round (R1-I8), the impact related to the use of chemical substances appears confusing (R2-I18). Expressly addressed in the comments by some experts, this lack of clarification is endorsed by the significant share of white votes ("I don't know") observed in the second round. Chemical substances hamper some recycling routes by decreasing the output value or by limiting end markets. Yet, it is not clear which exact chemical substances disrupt the process and to what extent. Classifications must be provided linking chemical substances to each specific technology, to advance alternatives in practice.

Refashion accordingly believes that the priority is to focus resources and effort on industrializing the recycling of non-reusable waste, before going all out into raising the collection rate.

4.2. Information

The Lack of Information on Materials Content

Essentially addressed from a technological perspective in the first round (R1-2 and R1-I5) the aspect was submitted distinctively in the second round to unravel opinions on current obstacles and became further explicit (R2-I3). Information availability on the product components is essential in directing to appropriate recycling routes. The current lack of information is considered as one of the critical challenges (65% positive votes and few votes against) in the sector (R2-I3) as labels are cut-off, inaccurate, or variations occur compared to the initial composition following the use phase (as highlighted by Wilting and Van Duijn in 2020 [27]). Therefore, improving identification of non-reusable textile flows is necessary to ensure homogeneous feedstock and enhance recycling applications.

4.3. Technologies

4.3.1. Importance of Technologies Supporting Identification

Overall, the current sorting process emerges as one of the major bottlenecks in the value chain (R1-I2, R1-I5, and R2-I6). The pre-sorting for quality aiming to differentiate reusable items from non-reusable ones will necessarily remain manual, but automated sorting is key to enhance material composition sorting and thus, recycling applications. However, the business model of the current sorting system is based on reuse categories towards second-hand markets and is not set up for composition sortation. Automated sorting machines supported by optical and artificial intelligence technologies are available (i.e., FIBERSORT, CETIA platform, SIPtex, or IMEC were mentioned), but the implementation is trivial to scale-up. Those technological infrastructures involve high investment costs, requiring enough demand for this sorted material and adaptation of the business model, which still needs to be resolved at this stage.

On the other hand, the lack of tracking technologies to convey information up to the end of life is usually identified as a critical challenge [13]. Still, the second round emphasizes a strong dissident among the experts (R2-I17). Quite a few consider the use of product-tracking technologies unrealistic, at least in the short term for mitigated experts, with regards to the fragmentation of the industry hindering data exchange across the value chain. It would not prevent inaccurate information or handle existing variations of content resulting from the use phase. Moreover, it would require the use of the same standards and would risk further impeding the process. Thus, some experts explicitly acknowledge optical sorting technologies as more promising. Opinions are very divided, and there is still a lack of evidence on the ability of tracking systems to overcome the consumer phase until the recycling stage in the sector. This contradicts previous research highlighting the requirement of such technologies to advance recycling (i.e., RFID et blockchain, notably) [9,14,32,46,62].

4.3.2. The Need for an Eased Disassembling Process

The manual dismantling stage prior to recycling is also recognized as a significant obstacle in Europe due to cost inefficiencies (R1-I6 and R2-I7). The iteration process between the two rounds enabled explicit opinions to be identified. While the deficiency of automated disassembling technologies still reaches positive consensus (65%) a persistent share of disagreement is discernible (R2-I7). Yet again, technologies are available (i.e., Picker from Laroche or CETIA platform mentioned) but technological improvement is needed. Some experts are divided on the potential leverage of these technologies, and instead address a design issue related to eco-design strategies. In contrast, the use of innovative manufacturing technologies (i.e., dissoluble yarns...) to smooth the dismantling process is prompting dubiousness (R1-I16/R2-I15).

4.3.3. The Lack of Scale-Up Innovative Recycling Technologies

The statements associated with the current state of recycling technologies nudged disagreements and raised several comments until the end of the study (R1-I17, R1-I18, and R2-I20). By decreasing the fiber length during the process, mechanical recycling limits closed-loop applications necessitating to mix short fibers with virgin ones to reach sufficient quality (as shown by the favorable consensus of 64% in the first round—R1-I17). The constraints associated with the process must be understood to make the best use of the technology. Still, it can provide diverse open-loop applications, and different markets must be considered to absorb all heterogenous feedstock (as reflected by the increasing share of mixed opinions in the second round—R2-I10). Progress for thermal and chemical recycling is underway to improve closed-loop opportunities. Several technologies are available for most material streams, but it is rather a challenge of investment and support to reach industrial scale (R1-I18 and R2-I12), which entails several critical aspects according to the comments (as shown by the disagreement in the second round—R2-I20). While addressed in the literature [14] the lack of environmental performance assessment of recycling technologies raised disagreement among the experts.

4.4. Markets

4.4.1. Insufficient Competitiveness of Recycled Content

The statements related to demand and economic viability of textile recycling stirred up strong disagreements among experts. The current state of demand for recycled textile materials is inconclusive. Still, a market shift is advanced with an increasing demand for textile-to-textile recycled products from brands. Several costs inefficiencies are resulting from the recycling pipeline, but opinion is divided between the low price of virgin materials and the lesser quality resulting from current mechanical recycling to explain the negligible share of closed-loop recycling (R2-I13).

4.4.2. The Lack of Investments and Long-Term Commitment

High investment costs and lack of long-term engagement are currently hindering the development of thermal and chemical recycling and the development of closed-loop recycling at this stage (R2-I12). One expert suggests a lack of economic, environmental, and technical scale-up vision, which discourages future investments. Overall, comments concur with the need for policies and collective actions to increase demand for recycled content.

4.4.3. The Need for Policies to Increase Market Opportunities

The absence of incentive policies to incorporate recycled content arises as the second most consensual statement in the second round (R2-I2) following several comments (R1-I3) and corroborates previous research addressing the need for regulations for the implementation of circular economy in the sector [12,51]. Public policies are needed to increase competitiveness of recycled content and support investments towards recycling technologies. More than incentives, the use of recycled content should be enforced by regulations according to some experts, and the recent Anti-Wastage and Circular Economy Law adopted in February 2020 in France was put forward.

4.5. Coordination

4.5.1. The Lack of Standards on Textile Recycling

At sector level, the lack of standards on notions such as "recyclability", "recycled content", or "disrupting factors" is failing to promote eco-design practices (R2-I4). More widely, further clarifications are needed to foster common understanding on textile recycling and disentangle impediments resulting from some disruptors in relation to existing recycling pathways. Yet, a few experts advised that rigid standards could hamper the constant innovation needed on textile recycling.

4.5.2. The Lack of Sector-Coordinated Vision

The lack of coordinated vision at the EU level fails to support the improvement of textile recycling (R2-I5), and a harmonized EU Textiles Strategy from the European Commission is expected. Throughout the study, several measures were mentioned to address either labelling issues or trade barriers, requiring defining up front priorities to enhance textile recycling.

4.5.3. The Deficient Collective Governance

The lack of collaboration and governance between the stakeholders along the value chain is sustained by a steady consensus (R1-I7 and R2-I9). Given the fragmentation of the industry and the correlated shortcomings along the value chain, a decisive organizational challenge is apparent. Comments range from the failure in governance with scattered initiatives, the lack of collaboration and specific actions, to the lack of communication. Altogether, coordination between the actors is crucial, and a few initiatives are mentioned, such as the Telaketju eco-system in Finland or the recent multi-stakeholder ReHubs project launched by EURATEX, aiming to set up an integrated system on five recycling hubs within Europe.

5. Limitations and Research Perspectives

5.1. Limitations

The first limitation of the study derives from the composition of the panel. Despite applying rigorous procedure rules, the selection inevitably entails shortcomings. Due to the wide range of statements and stakeholder groups, expertise was unevenly distributed among the experts, and it likely influenced the results. However, the panel was coherent with regards to the research objective, which did not aim to provide an in-depth exposition of the topic. Moreover, the study did not aim to explore the consumer's perspective. This limitation inevitably narrows the analysis of some barriers, such as the lack of awareness or the lack of demand for recycled content, requiring further investigations on consumer behaviors [79]. To generate valid results, the study required us to involve high-skilled experts with a substantial experience in the field. However, the opinion of younger generations of professionals should be considered in future research [80] especially to explore enablers to unravel these textile recycling challenges.

Inherent to the Delphi method used, the second limitation is the design of the first questionnaire. Although experts had the possibility to submit comments and propose other statements later on, the orientation of the first questionnaire from the start probably created a subjective bias.

Another limitation results from a diverging interpretation observed on certain terms. We noticed that "textile recycling" was sometimes comprehended as equivalent to "textile-to-textile recycling". Although the wording of the sentence was carefully reviewed to limit the risk of misinterpretation, it certainly influenced the results. Still, experts were asked to provide comments in support of their votes, which enabled us to limit this shortcoming through a cautious review process.

Finally, a last shortcoming resides in the research approach applied. While the Delphi-Régnier method allowed us to gather empirical data and visualize expert's opinions to identify the priority challenges of textile recycling, further statistical analysis should be conducted to corroborate the results and to define precise roadmaps for the sector.

5.2. Research Perspectives

Following this research, several perspectives have emerged.

From a theoretical perspective, only a few comprehensive frameworks on textile recycling were found in the extant literature. These frameworks mainly consist in mappings of the different stages of the value chain. Consequently, a significant obstacle was encountered when attempting to delimit the scope of the review.

The paper highlighted a lack of empirical knowledge in the literature. Based on the findings, several case studies could be conducted, allowing deeper level of observation and in-depth results. Case studies could investigate the implementation of traceability solutions in the textile value chain up to the recycling stage, as the technical, economic, and organizational viability of such solutions is still unascertained. Especially, in view of the apparent dissidence, it could advance research on circularity in the industry by investigating the leverage of these technologies in closing the textile loop. Case studies exploring stakeholder projects, such as the Rehubs initiative launched by EURATEX, could provide valuable insights on the governance and collaborative challenges necessitating by such innovative and large-scale management of material streams. Further case studies could research innovative business models to explore the implementation of circular strategies in the industry, especially with regards to the tensions between durability and recyclability. Given the lack of consensus on the demand for recycled products, more research should also be conducted with consumers to explore their perception regarding recycled products [79].

Finally, our study is qualitative and thus only elicits the main consensus among the experts on the existing challenges in the sector. It does not enable to prioritize the obstacles towards a fine roadmap or to suggest interconnections between the different critical aspects. Nevertheless, the findings show evident opportunities for extra research. Other methodologies could be used, such as the multiple-criteria decision analysis (MCDA) or the econometric analysis, to corroborate, invalidate, and further clarify the findings of the study.

6. Conclusions

This paper aimed to deepen understanding of the current challenges of textile recycling and the priority concerns to address in the fashion and apparel sector. To achieve these research objectives, a systematic literature review supplemented by a Delphi-Régnier study was performed.

Regarding the first research question, the coding process applied during the literature review allowed us to discern a set of challenges in the recycling value chain. Through the iterative process of the Delphi-Régnier study, those challenges were discussed and revised, based on experts' comments between the two rounds, advancing empirical knowledge on the topic. Based on the findings, the lack of eco-design practices emerges consensually as the major barrier preventing the enhancement of textile recycling. Innovative recycling technologies are crucial to develop textile-to-textile recycling and are now available for the main material streams. At this stage, industrialization is rather problematic and is interrelated with several other critical aspects. Beyond requiring substantial investments, recycling technologies entail abundant and suitable textile flows input to scale-up. Yet, the current manual sorting system arises as a significant bottleneck in the value chain, due to discrepancies with the paramount separation for material composition towards recycling applications. The lack of information availability and accuracy on products components hinders directing textile feedstocks to appropriate recycling routes. Hence, improving textile recycling is closely related to the identification of technologies. As one of the main original findings, a strong dissidence prevails on the potential of product-tracking technologies to convey information up to downstream in the value chain, and experts are more inclined to automated sorting technologies. Still, the business case needs to be resolved as the entire management system of used textiles flows is rationally structured around reusability, in accordance with the waste hierarchy. Consequently, more than a technical issue, the study instead suggests an essential restructuring of this complex value chain to support the improvement of this resource management system, highlighted by the strong focus on challenges at the sector level.

Through the emergence of consensus and dissensus on the topic, we were able to answer the second research question. The lack of eco-design practices, the absence of incentive policies, and the lack of available and accurate information on the product components emerge as the most consensual statements in the second round. Based on the results obtained from this iterative process, three dominant challenges distinctively stand out, without specific order of importance: (1) The need to advance eco-design practices consistently with respect to reusability and recyclability strategies; (2) The strong requisite of policies and governance to advance textile recycling; (3) The necessity to unravel the deficiencies related to the current manual sorting system.

Based on a European perspective, the study significantly emphasizes the critical impact of public policies and provides support for several recommendations to enhance textile recycling in the sector. Especially, in the perspective of the mandatory separate collection [41], the findings call for measures endorsing the optimization and industrialization of the recycling value chain, rather than an increased collection of used textiles in Europe. Finally, by refining some general assertions on the topic and providing extensive empirical knowledge on the topic, this paper can help practitioners to advance relevant initiatives. The study also highlights several research opportunities, especially on the leverage of traceability technologies to enhance textile recycling in the industry. However, the method used has inherent limitations, and further statistical analysis should be conducted to further explore the priorities or interrelations between such challenges in the sector.

Author Contributions: Conceptualization, J.R., A.-A.L.; methodology, J.R. and S.L.; writing—original draft preparation, J.R.; writing—review and editing, S.L. and L.G. and A.-A.L. All authors have read and agreed to the published version of the manuscript.

Funding: This research received no external funding.

Institutional Review Board Statement: Not applicable.

Informed Consent Statement: Not applicable.

Data Availability Statement: The data presented in this study are available on request from the corresponding author due to privacy.

Acknowledgments: We would like to thank the 28 experts and their organizations for agreeing to participate in the study by taking the time and involved themself by sharing their expertise to contribute to scientific knowledge.

Conflicts of Interest: The authors declare no conflict of interest.

Appendix A

Figure A1. Overview of the ColorInsight website used to submit the questionnaires to the experts.

Appendix B

First Questionnaire Statements with Their Vote's Distribution

| 7% | 50% | 18% | 18% | |

The lack of dedicated and separate collection infrastructures in several countries for textile waste prevents the development of textile recycling.

| 43% | 25% | 29% | | |

The lack of efficient and cost-effective automated sorting technologies, able to provide well-defined and homogeneous feedstock (composition, quality and color) is a major barrier to enhance textile recycling.

| 18% | 46% | 25% | 11% | |

Enhancing recycling further requires advanced information systems to exchange data across the value chain on the precise composition and chemical inputs of discarded textiles.

| 29% | 11% | 43% | 7% | 11% |

The lack of innovative manufacturing technologies prevents eased disassembly of the product at the end-of-life.

| 32% | 25% | 25% | 18% | |

The absence of technologies enabling an eased disassembling process hinders the development of textile recycling.

| 11% | 36% | 36% | 11% | |

The current limited recycling technologies prevent the development of textile recycling towards high-value applications.

| 25% | 29% | 18% | 14% | 7% | |

The development of textile recycling is restrained by the insufficient maturity of new recycling technologies

| 21% | 32% | 21% | 14% | 7% |

The insufficient connection between demand and supply of textile waste inhibits resource availability and effective trading of recycled materials.

| 57% | 18% | 18% | | |

The large variety of blended materials in the industry is a major barrier as reducing opportunities for textile recycling.

| 25% | 11% | 32% | 25% | 7% |

The construction of the yarn or the fabric is an important potential disruptor to the recycling process.

| 14% | 29% | 21% | 21% | 11% |

The increasing introduction of new materials in the industry is a potential inhibiting factor by creating challenges with recycling streams.

| 21% | 36% | 36% | | |

The use of chemical substances contribute to disrupt the recycling process by decreasing the value of the output.

| 29% | 46% | 14% | 7% | |

The presence of external accessories (such as buttons, zipper, carelabels..) is another important disruptor of the recycling process.

| 25% | 21% | 32% | 7% | 11% |

The washing and wearing conditions degrade the fibers and hampers the sorting and recycling process affecting the recycled feedstock quality.

| 21% | 43% | 7% | 14% | 11% |

The available mechanical recycling technologies shortens the fibers during the process, preventing to provide a wide range of qualitative recycled output.

| 11% | 32% | 14% | 25% | 14% |

The available mechanical recycling technologies only provides outputs with limited recyclability.

| 43% | 11% | 14% | 25% | 7% |

The limited profitability of textile recycling is a significant bottleneck in textile recycling, preventing recycled materials to compete with virgin ones.

| 29% | 25% | 18% | 14% | 7% | |

The high costs of research and development associated with the development of new recycling technologies discourages prospective investments.

| 25% | 18% | 7% | 32% | 11% | 7% |

The development of textile recycling is restrained by the uncertain end-market opportunities.

| 25% | 25% | 32% | 11% | |

The development of textile recycling is restrained by a limited public awareness, implying low collection rates.

| 25% | 39% | 21% | 11% | |

The lack of collaboration and alignment between all the stakeholders (collectors, sorters, recyclors, manufacturers, brands, policy-makers, consumers) curbs the improvement of textile recycling.

| 43% | 29% | 18% | 7% |

The missing clear regulatory framework from policy-makers fails to support the improvement of textile recycling.

| 11% | 29% | 36% | 14% | 7% |

The lack of accurate demonstration of environmental performance for innovative recycling technologies and the overall end-of-life treatment is a challenge in textile recycling.

Appendix C

Second Questionnaire Statements with Their Vote's Distribution

| 30% | 17% | 17% | 17% | 13% |

EDITED: No separation at collection points (e.g., mix of textiles with different compositions...) is a barrier to recycling value chain efficiency

| 30% | 35% | 26% | 9% |

EDITED: The lack of available and accurate information on material and chemical content prevents access to appropriate recycling channels

| 22% | 43% | 26% |

EDITED: The lack of automated identification and sorting technologies (such as optical sorting and artifical intelligence technologies) is a barrier to recycling value chain efficiency

| 17% | 17% | 35% | 17% | 9% |

EDITED: The use of product tracking technologies (e.g., digital passports) by fashion companies seems unrealistic to ensure information traceability on composition and chemical substances until end-of-life

| 17% | 26% | 35% | 13% |

EDITED: The lack of innovative manufacturing solutions prevents performant disassembling (e.g., use of disoluble yarns by fashion companies etc...)

| 22% | 43% | 17% | 17% |

EDITED: The deficiency of automated disassembling technologies (meaning separation technologies prior to recycling) hinders recycling cost-efficiency

| 48% | 22% | 13% | 9% | 9% |

EDITED: Failure to identify appropriate feedstocks through digital platforms is limiting open-loop recycling opportunities

| 35% | 39% | 9% | 9% |

EDITED: The current profusion of material mix and fiber blends in the industry reduces the market opportunities

| 13% | 43% | 17% | 17% |

EDITED: The presence of external accessories (e.g., buttons, zippers...) is a major barrier to recycling value chain efficiency

| 13% | 35% | 26% | 9% | 13% |

EDITED: The current presence of chemical substances in textiles limits market applications, specifically by hampering chemical recycling process

| 22% | 35% | 30% | 9% |

EDITED: The low quality of post-consumer feedstock impede closed-loop recovery

| 39% | 43% | 13% |

NEW ITEM: The lack of eco-design practices on product recyclability by fashion companies is a major obstacle to develop textile recycling

| 17% | 35% | 39% | 9% |

EDITED: The reduction of fiber length during the mechanical recycling process brings limitation to end-market opportunities

| 9% | 22% | 39% | 9% | 9% | 13% |

EDITED: The scarcity of available recycled products is due to the lack of scaled-up thermal and chemical recycling technologies

| 35% | 13% | 22% | 17% | 9% |

EDITED: Recycled content is not competitive enough compared to low priced virgin materials for garment applications

| 39% | 22% | 13% | 9% | 17% |

EDITED: The high costs of investment and difficulties to commit long-term for scale-up hinders the development of thermal and chemical recycling technologies

| 26% | 39% | 22% |

EDITED: The lack of communication and governance between all the value chain actors obstructs the improvement of textile recycling

| 22% | 48% | 22% |

EDITED: The lack of a coordinated vision at the EU level fails to support the improvement of textile recycling

| 30% | 35% | 26% |

NEW ITEM: The absence of textile recycling standards (such as: recyclability, recycled content, disruptors to recycling...) fails to promote eco-design practices

39%	30%	22%		

NEW ITEM: Absence of incentive policies to incorporate recycled content is limiting the demand for recycled materials

26%	13%	22%	9%	26%

EDITED: The lack of efficient way to compare the recycling process between them does not promote the use of recycled materials

Appendix D

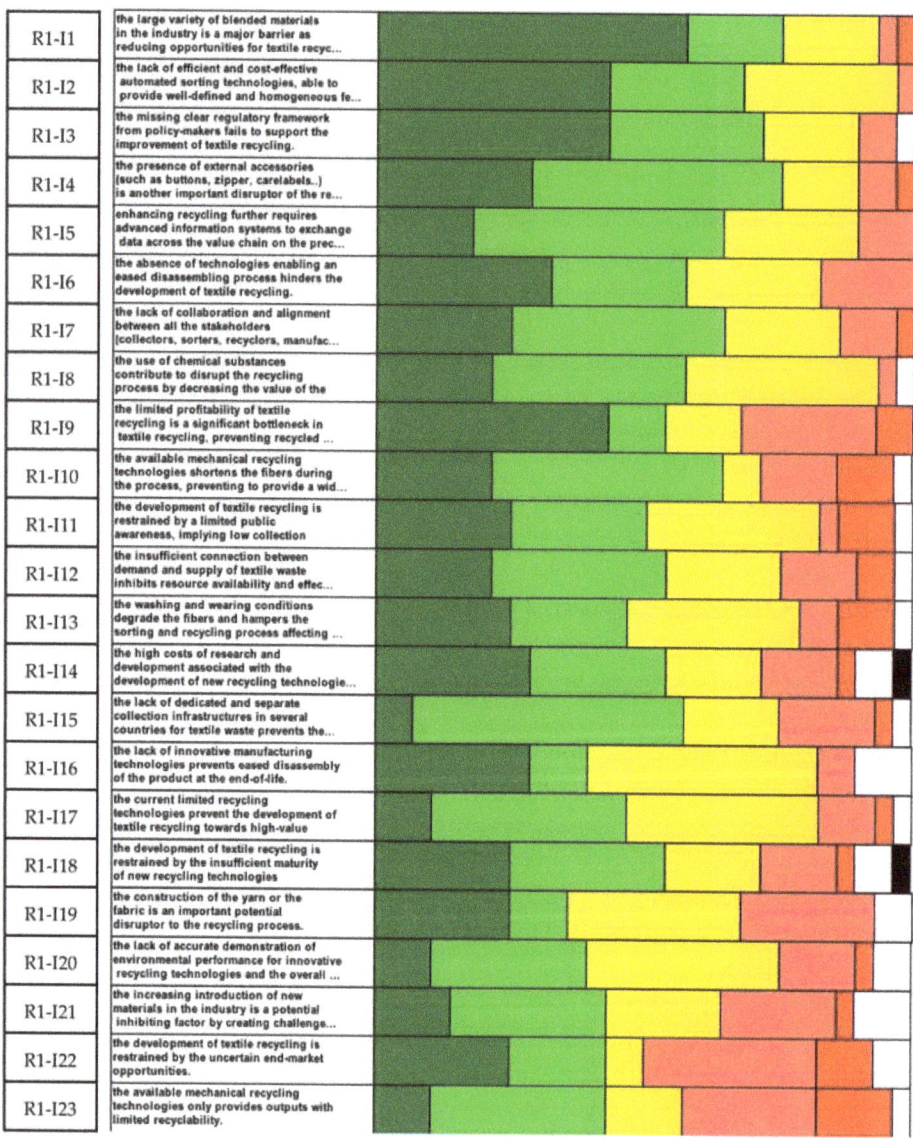

Figure A2. Items matrix of the first questionnaire (first round).

Appendix E

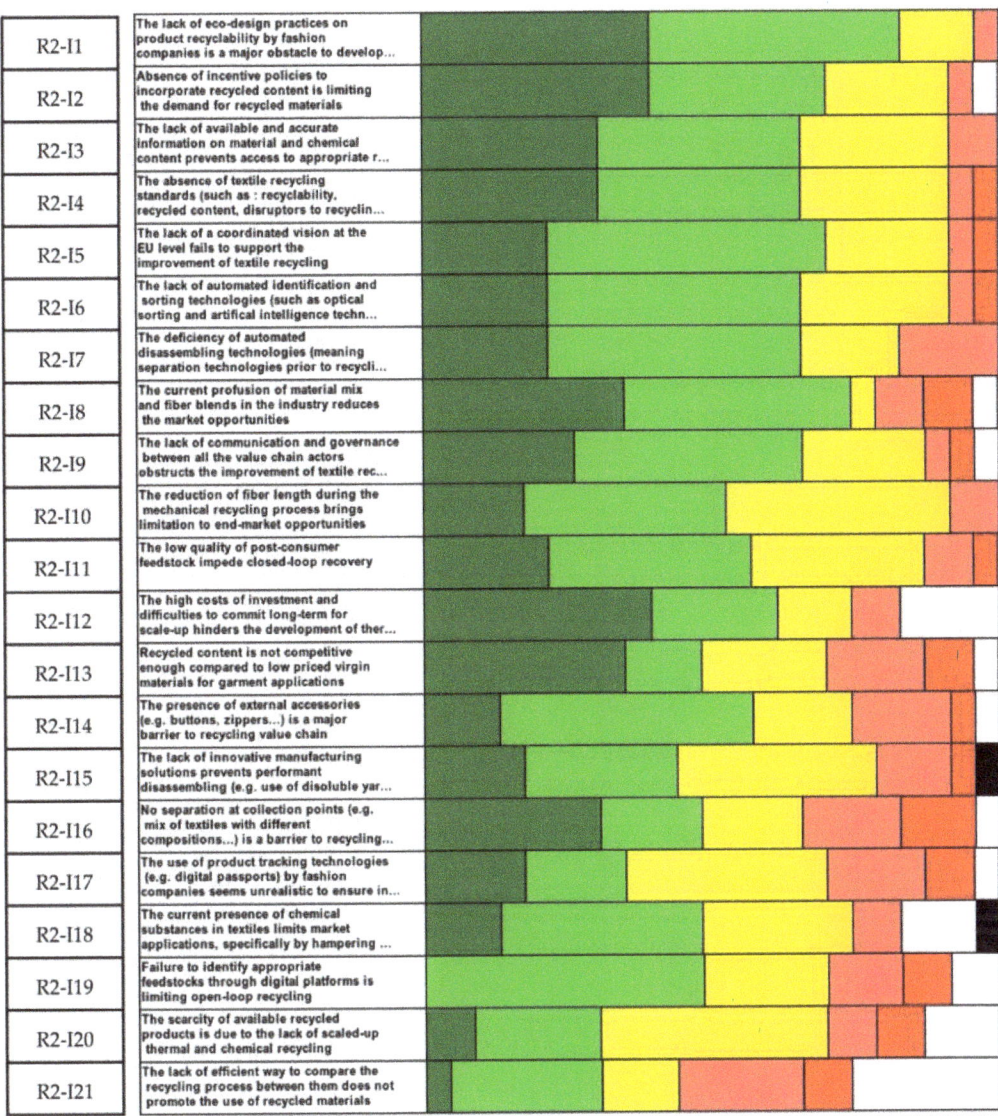

Figure A3. Items matrix of the second questionnaire (second round).

References

1. Statista. Global Apparel Market—Statistics & Facts | Statista. Available online: https://www.statista.com/topics/5091/apparel-market-worldwide/ (accessed on 25 August 2021).
2. Gereffi, G.; Frederick, S. *The Global Apparel Value Chain, Trade, and the Crisis: Challenges and Opportunities for Developing Countries*; Policy Research Working Paper 5281; World Bank: Washington, DC, USA, 2010. Available online: https://openknowledge.worldbank.org/handle/10986/3769 (accessed on 23 May 2021).
3. Niinimäki, K.; Peters, G.; Dahlbo, H.; Perry, P.; Rissanen, T.; Gwilt, A. The Environmental Price of Fast Fashion. *Nat. Rev. Earth Environ.* **2020**, 2, 189–200. [CrossRef]

4. Jacomet, D.; Minvielle, G. De-Industrialisation Re-Industrialisation in the Fashion Industry. Available online: https://www.ifmparis.fr/fr/recherche-academique/desindustrialisation-reindustrialisation-dans-lindustrie-de-la-mode (accessed on 23 May 2021).
5. European Commission. International Trade | Internal Market, Industry, Entrepreneurship and SMEs. Available online: https://ec.europa.eu/growth/sectors/fashion/textiles-and-clothing-industries/international-trade_en (accessed on 25 August 2021).
6. Kumar, V.; Agrawal, T.K.; Wang, L.; Chen, Y. Contribution of Traceability towards Attaining Sustainability in the Textile Sector. *Text. Cloth. Sustain.* **2017**, *2*, 5. [CrossRef]
7. McKinsey & Company; Global Fashion Agenda. Fashion on Climate: How the Industry Can Urgently Act to Reduce Its Greenhouse Emissions. 2020. Available online: https://www.mckinsey.com/industries/retail/our-insights/fashion-on-climate (accessed on 23 May 2021).
8. Quantis. Measuring Fashion Environmental Impact of the Global Apparel and Footwear Industries: Environmental Impact of the Global Apparel and Footwear Industries. 2018. Available online: https://quantis-intl.com/report/measuring-fashion-report (accessed on 23 May 2021).
9. Ellen MacArthur Foundation. A New Textiles Economy: Redesigning Fashion's Future. 2017. Available online: https://ellenmacarthurfoundation.org/a-new-textiles-economy (accessed on 4 October 2021).
10. Quantis. Draft Product Environmental Footprint—Representative Product (PEF-RP) Study Report Apparel and Footwear, Version 1.2. PEFCR for Apparel and Footwear—Stakeholder Workspace—EU Environmental Footprint—EC Extranet Wiki. 2021. Available online: https://europa.eu (accessed on 7 July 2021).
11. United Nations Environment Programme. Sustainability and Circularity in the Textile Value Chain. 2020. Available online: https://wedocs.unep.org/handle/20.500.11822/34184 (accessed on 10 March 2021).
12. Brydges, T. Closing the Loop on Take, Make, Waste: Investigating Circular Economy Practices in the Swedish Fashion Industry. *J. Clean. Prod.* **2021**, *293*, 126245. [CrossRef]
13. Sandvik, I.M.; Stubbs, W. Circular Fashion Supply Chain through Textile-to-Textile Recycling. *J. Fash. Mark. Manag. Int. J.* **2019**, *2*, 366–381. [CrossRef]
14. Allwood, J.M. *Well Dressed? The Present and Future Sustainability of Clothing and Textiles in the United Kingdom*; University of Cambridge Institute for Manufacturing: Cambridge, UK, 2006. Available online: https://www.researchgate.net/publication/282249347_Well_Dressed_The_Present_and_Future_Sustainability_of_Clothing_and_Textiles_in_the_United_Kingdom (accessed on 29 September 2021).
15. D'Adamo, I.; Lupi, G. Sustainability and Resilience after COVID-19: A Circular Premium in the Fashion Industry. *Sustainability* **2021**, *2*, 1861. [CrossRef]
16. Koszewska, M. Circular Economy—Challenges for the Textile and Clothing Industry. *Autex Res. J.* **2018**, *2*, 337–347. [CrossRef]
17. Roos, S.; Sandin, G.; Peters, G.; Spak, B.; Schwarz Bour, L.; Perzon, E.; Jönsson, C. *White Paper on Textile Recycling*; Mistra Future Fashion: Stockholm, Sweden, 2019.
18. Dahlbo, H.; Aalto, K.; Eskelinen, H.; Salmenperä, H. Increasing Textile Circulation—Consequences and Requirements. *Sustain. Prod. Consum.* **2017**, *9*, 44–57. [CrossRef]
19. Sajn, N. *Environmental Impact of the Textile and Clothing Industry: What Consumers Need to Know*; European Parliament: Brussels, Belgium, 2019. Available online: https://www.europarl.europa.eu/thinktank/en/document.html?reference=EPRS_BRI%282019%29633143 (accessed on 7 July 2021).
20. European Environment Agency. Textiles in Europe's Circular Economy. 2019. Available online: https://www.eea.europa.eu/publications/textiles-in-europes-circular-economy (accessed on 7 July 2021).
21. Intergovernmental Panel on Climate Change (IPCC). Climate Change 2021 the Physical Science Basis: Summary for Policymakers. 2021. Available online: https://www.ipcc.ch/report/ar6/wg1/ (accessed on 7 July 2021).
22. Bell, J.E.; Autry, C.W.; Mollenkopf, D.A.; Thornton, L.M. A Natural Resource Scarcity Typology: Theoretical Foundations and Strategic Implications for Supply Chain Management. *J. Bus. Logist.* **2012**, *2*, 158–166. [CrossRef]
23. Geissdoerfer, M.; Savaget, P.; Bocken, N.M.P.; Hultink, E.J. The Circular Economy—A New Sustainability Paradigm? *J. Clean. Prod.* **2017**, *143*, 757–768. [CrossRef]
24. McDonough, W.; Braungart, M. *Cradle to Cradle: Remaking the Way We Make Things*; North Point Press: New York, NY, USA, 2010.
25. Ellen MacArthur Foundation. Towards the Circular Economy: Opportunities for the Goods Sector. 2013. Available online: https://emf.thirdlight.com/link/coj8yt1jogq8-hkhkq2/@/preview/1?o (accessed on 23 May 2021).
26. Colucci, M.; Vecchi, A. Close the Loop: Evidence on the Implementation of the Circular Economy from the Italian Fashion Industry. *Bus. Strategy Environ.* **2021**, *2*, 856–873. [CrossRef]
27. Harmsen, P.; Scheffer, M.; Bos, H. Textiles for Circular Fashion: The Logic behind Recycling Options. *Sustainability* **2021**, *2*, 9714. [CrossRef]
28. Rosa, P.; Sassanelli, C.; Terzi, S. Towards Circular Business Models: A Systematic Literature Review on Classification Frameworks and Archetypes. *J. Clean. Prod.* **2019**, *236*, 117696. [CrossRef]
29. Chen, W.-K.; Nalluri, V.; Hung, H.-C.; Chang, M.-C.; Lin, C.-T. Apply DEMATEL to Analyzing Key Barriers to Implementing the Circular Economy: An Application for the Textile Sector. *Appl. Sci.* **2021**, *2*, 3335. [CrossRef]

30. Niinimaki, K. *Sustainable Fashion in a Circular Economy*; Niinimaki, K., Ed.; Aalto University School of Arts, Design and Architecture: Espoo, Finland, 2019; pp. 1–249. Available online: https://shop.aalto.fi/media/filer_public/53/dc/53dc45bd-9e9e-4d83-916d-1d1ff6bf88d2/sustainable_fashion_in_a_circular_economyfinal.pdf (accessed on 23 May 2021).
31. Bressanelli, G.; Saccani, N.; Perona, M.; Baccanelli, I. Towards Circular Economy in the Household Appliance Industry: An Overview of Cases. *Resources* **2020**, *2*, 128. [CrossRef]
32. Kazancoglu, I.; Kazancoglu, Y.; Yarimoglu, E.; Kahraman, A. A Conceptual Framework for Barriers of Circular Supply Chains for Sustainability in the Textile Industry. *Sustain. Dev.* **2020**, *2*, 1477–1492. [CrossRef]
33. Potting, J.; Hekkert, M.; Worrell, E.; Hanemaaijer, A. *Circular Economy: Measuring Innovation in The Product Chain*; PBL Publishers: The Hague, The Netherlands, 2017. Available online: https://dspace.library.uu.nl/handle/1874/358310 (accessed on 28 January 2021).
34. European Commission. Waste Framework Directive. Available online: https://ec.europa.eu/environment/topics/waste-and-recycling/waste-framework-directive_en (accessed on 3 August 2021).
35. Sandin, G.; Peters, G.M. Environmental Impact of Textile Reuse and Recycling—A Review. *J. Clean. Prod.* **2018**, *184*, 353–365. [CrossRef]
36. Shirvanimoghaddam, K.; Motamed, B.; Ramakrishna, S.; Naebe, M. Death by Waste: Fashion and Textile Circular Economy Case. *Sci. Total Environ.* **2020**, *718*, 137317. [CrossRef]
37. Peters, G.M.; Sandin, G.; Spak, B. Environmental Prospects for Mixed Textile Recycling in Sweden. *ACS Sustain. Chem. Eng.* **2019**, *2*, 11682–11690. [CrossRef]
38. Payne, A. Open- and Closed-Loop Recycling of Textile and Apparel Products. In *Handbook of Life Cycle Assessment (LCA) of Textiles and Clothing*; Elsevier: Amsterdam, The Netherlands, 2015; pp. 103–123.
39. Hole, G.; Hole, A.S. Improving Recycling of Textiles Based on Lessons from Policies for Other Recyclable Materials: A Minireview. *Sustain. Prod. Consum.* **2020**, *23*, 42–51. [CrossRef]
40. European Commission. Green Deal. Un Pacte Vert pour l'Europe | Commission Européenne. Available online: https://ec.europa.eu (accessed on 5 August 2021).
41. European Commission. EU Strategy for Sustainable Textiles. Available online: https://ec.europa.eu/info/law/better-regulation/have-your-say/initiatives/12822-EU-strategy-for-sustainable-textiles (accessed on 5 August 2021).
42. European Commission. Sustainable Products Initiative. Available online: https://ec.europa.eu/info/law/better-regulation/have-your-say/initiatives/12567-Sustainable-products-initiative (accessed on 5 August 2021).
43. Hsu, C.-C.; Sandford, B.A. The Delphi Technique: Making Sense of Consensus. *Pract. Assess. Res. Eval.* **2007**, *2*, 1–8.
44. Régnier, F. L'Abaque de Régnier: Pour une lecture musicale de la communication: Exemple d'application en éducation permanente. *Acta Endosc.* **1983**, *13*, 361–373. [CrossRef]
45. Tranfield, D.; Denyer, D.; Smart, P. Towards a Methodology for Developing Evidence-Informed Management Knowledge by Means of Systematic Review. *Br. J. Manag.* **2003**, *2*, 207–222. [CrossRef]
46. Leal Filho, W.; Ellams, D.; Han, S.; Tyler, D.; Boiten, V.J.; Paço, A.; Moora, H.; Balogun, A.-L. A Review of the Socio-Economic Advantages of Textile Recycling. *J. Clean. Prod.* **2019**, *218*, 10–20. [CrossRef]
47. Rotimi, E.O.O.; Topple, C.; Hopkins, J. Towards A Conceptual Framework of Sustainable Practices of Post-Consumer Textile Waste at Garment End of Lifecycle: A Systematic Literature Review Approach. *Sustainability* **2021**, *2*, 2965. [CrossRef]
48. Cura, K.; Rintala, N.; Kamppuri, T.; Saarimäki, E.; Heikkilä, P. Textile Recognition and Sorting for Recycling at an Automated Line Using Near Infrared Spectroscopy. *Recycling* **2021**, *2*, 11. [CrossRef]
49. Palacios-Mateo, C.; van der Meer, Y.; Seide, G. Analysis of the Polyester Clothing Value Chain to Identify Key Intervention Points for Sustainability. *Environ. Sci. Eur.* **2021**, *2*, 2. [CrossRef] [PubMed]
50. Riba, J.-R.; Cantero, R.; Canals, T.; Puig, R. Circular Economy of Post-Consumer Textile Waste: Classification through Infrared Spectroscopy. *J. Clean. Prod.* **2020**, *272*, 123011. [CrossRef]
51. Jia, F.; Yin, S.; Chen, L.; Chen, X. The Circular Economy in the Textile and Apparel Industry: A Systematic Literature Review. *J. Clean. Prod.* **2020**, *259*, 120728. [CrossRef]
52. Piribauer, B.; Bartl, A. Textile Recycling Processes, State of the Art and Current Developments: A Mini Review. *Waste Manag. Res. J. Sustain. Circ. Econ.* **2019**, *2*, 112–119. [CrossRef] [PubMed]
53. Ütebay, B.; Çelik, P.; Çay, A. Effects of Cotton Textile Waste Properties on Recycled Fibre Quality. *J. Clean. Prod.* **2019**, *222*, 29–35. [CrossRef]
54. Nørup, N.; Pihl, K.; Damgaard, A.; Scheutz, C. Evaluation of a European Textile Sorting Centre: Material Flow Analysis and Life Cycle Inventory. *Resour. Conserv. Recycl.* **2019**, *143*, 310–319. [CrossRef]
55. Bukhari, M.A.; Carrasco-Gallego, R.; Ponce-Cueto, E. Developing a National Programme for Textiles and Clothing Recovery. *Waste Manag. Res. J. Sustain. Circ. Econ.* **2018**, *2*, 321–331. [CrossRef] [PubMed]
56. European Parliament. Directive 2008/98/EC of the European Parliament and of the Council of 19 November 2008 on waste and repealing certain Directives. *J. Eur. Union* **2008**, *312*, 312–530.
57. Hawley, J.M. Textile Recycling: A System Perspective. In *Recycling in Textiles*; Elsevier: Amsterdam, The Netherlands, 2006; pp. 7–24.
58. Leonas, K.K. The Use of Recycled Fibers in Fashion and Home Products. In *Textiles and Clothing Sustainability*; Muthu, S.S., Ed.; Textile Science and Clothing Technology; Springer: Singapore, 2017; pp. 55–77.

59. Reverse Resources. The Undiscovered Business Potential of Production Leftovers within Global Fashion Supply Chains: Creating a Digitally Enhanced Circular Economy. 2017. Available online: https://reverseresources.net/news/white-paper-by-rr (accessed on 10 March 2021).
60. Palm, D. Towards a Nordic Textile Strategy: Collection, Sorting, Reuse and Recycling of Textiles; Nordic Council of Ministers. 2014. Available online: https://norden.diva-portal.org/smash/get/diva2:720964/FULLTEXT01.pdf (accessed on 23 May 2021).
61. Re_Fashion. Annual Report. Available online: https://refashion.fr/pro/sites/default/files/rapport-etude/refashion-annual-report-2020.pdf (accessed on 15 August 2021).
62. Fibersort Interreg Northwest Europe. Overcoming Barriers for Long-Term Implementation. 2020. Available online: https://www.circle-economy.com/resources/fibersort-overcoming-barriers-for-long-term-implementation (accessed on 23 May 2021).
63. Elander, M.; Ljungkvist, A. *Critical Aspects in Design for Fiber-to-Fiber Recycling of Textiles*; IVL Swedish Environmental Research Institute Ltd.: Stockholm, Sweden, 2016. Available online: http://mistrafuturefashion.com/wp-content/uploads/2016/06/MFF-report-2016-1-Critical-aspects.pdf (accessed on 23 May 2021).
64. Zimmermann, W. Biocatalytic Recycling of Polyethylene Terephthalate Plastic. *Philos. Trans. R. Soc. Math. Phys. Eng. Sci.* **2020**, *2*, 20190273. [CrossRef]
65. Pensupa, N.; Leu, S.-Y.; Hu, Y.; Du, C.; Liu, H.; Jing, H.; Wang, H.; Lin, C.S.K. Recent Trends in Sustainable Textile Waste Recycling Methods: Current Situation and Future Prospects. *Top. Curr. Chem.* **2017**, *2*, 76. [CrossRef]
66. Textile Exchange Preferred Fiber and Materials Market Report 2020. Textile Exchange. 2020. Available online: https://textileexchange.org/2020-preferred-fiber-and-materials-market-report-pfmr-released-2 (accessed on 28 May 2021).
67. Awaja, F.; Pavel, D. Recycling of PET. *Eur. Polym. J.* **2005**, *2*, 1453–1477. [CrossRef]
68. Wilting, J.; Hilde, V.D. Clothing Labels: Accurate or Not? *Circle Econ.* 2020. Available online: https://www.circle-economy.com/resources/clothing-labels-accurate-or-not (accessed on 15 January 2021).
69. Dalkey, N.; Helmer, O. An Experimental Application of the DELPHI Method to the Use of Experts. *Manag. Sci.* **1963**, *2*, 458–467. [CrossRef]
70. Gordon, T.J. The Delphi Method. In *Future Research Methodology—Version 3.0*; AC/UNU Millennium Project: Washington, DC, USA, 1994.
71. Gardas, B.B.; Raut, R.D.; Narkhede, B. Modelling the Challenges to Sustainability in the Textile and Apparel (T&A) Sector: A Delphi-DEMATEL Approach. *Sustain. Prod. Consum.* **2018**, *15*, 96–108.
72. Agrawal, T.; Pal, R. Traceability in Textile and Clothing Supply Chains: Classifying Implementation Factors and Information Sets via Delphi Study. *Sustainability* **2019**, *2*, 1698. [CrossRef]
73. Garcia-Torres, S.; Rey-Garcia, M.; Sáenz, J.; Seuring, S. Traceability and Transparency for Sustainable Fashion-Apparel Supply Chains. *J. Fash. Mark. Manag. Int. J.* **2021**, in press. [CrossRef]
74. Delbecq, A.L.; Van de Ven, A.H.; Gustafson, D.H. *Group Techniques for Program Planning: A Guide to Nominal Group and Delphi Processes*; Management Applications Series; Scott Foresman: Glenview, IL, USA, 1975.
75. Moeuf, A.; Lamouri, S.; Pellerin, R.; Tamayo-Giraldo, S.; Tobon-Valencia, E.; Eburdy, R. Identification of Critical Success Factors, Risks and Opportunities of Industry 4.0 in SMEs. *Int. J. Prod. Res.* **2020**, *2*, 1384–1400. [CrossRef]
76. Color Insight. Available online: https://www.colorinsight.fr/ (accessed on 5 February 2021).
77. Ashton, R.H. Combining the Judgments of Experts: How Many and Which Ones? *Organ. Behav. Hum. Decis. Process.* **1986**, *2*, 405–414. [CrossRef]
78. Jones, C.G. A Delphi Evaluation of Agreement between Organizations. In *The Delphi Method: Techniques and Applications*; Addison-Wesley: Boston, MA, USA, 1975; pp. 160–167.
79. Djafarova, E.; Bowes, T. 'Instagram Made Me Buy It': Generation Z Impulse Purchases in Fashion Industry. *J. Retail. Consum. Serv.* **2021**, *59*, 102345. [CrossRef]
80. Appolloni, A.; D'Adamo, I.; Gastaldi, M.; Santibanez-Gonzalez, E.D.R.; Settembre-Blundo, D. Growing E-Waste Management Risk Awareness Points towards New Recycling Scenarios: The View of the Big Four's Youngest Consultants. *Environ. Technol. Innov.* **2021**, *23*, 101716. [CrossRef]

Article

Three-Dimensional Printing Fashion Product Design with Emotional Durability Based on Korean Aesthetics

Seonju Kam

Department of Clothing and Textiles, Kyung Hee University, Seoul 02447, Korea; sjkam@khu.ac.kr;
Tel.: +82-70-8610-2255

Abstract: Given the potentially significant environmental impacts of fashion design, various design approaches are required to extend product lifespan. Digital design methods may play an essential role in reducing the environmental impact of products and production processes. In addition, a design approach inspired by nature, where humans have long lived, is valid for sustainable design innovation. The purpose of this study is to examine the aesthetics of Koreans, who prefer nature, and to find a sustainable fashion design approach by using it as a knowledge database. In this study, a parametric design methodology that can reflect knowledge-based data in the process of producing 3D printing sustainable fashion products, considering the emotional durability of consumers, was used. The study results are as follows. From the aesthetic point of view of Korea, sustainable design characteristics represent unique Korean folk art, resilience to nature, and simplicity that resembles nature. The properties of the form represented to "forms resembling nature", "changeable forms", "organic forms", and "minimal forms". Materials were "nature inspired textures", "rustic natural materials", and "regional materials". Colors were "the colors of nature" and "indigenous colors". The parametric controls variables used for 3D printing the fashion products were size, assembly style, and sustainable material. These control parameters were used to create designs according to the individual taste of users. In the 3D printing fashion product design process, pieces were printed in different shapes and sizes by controlling the parameters to create designs according to users' tastes and Korean aesthetics. It was determined that this process could extend the lifespan of products, and that it is possible to modify sustainable fashion products according to personal taste by adjusting numerical values and extracting visual images based on knowledge of art and culture.

Keywords: emotional durability; 3D printing fashion product design; Korean aesthetic

Citation: Kam, S. Three-Dimensional Printing Fashion Product Design with Emotional Durability Based on Korean Aesthetics. *Sustainability* 2022, 14, 240. https://doi.org/10.3390/su14010240

Academic Editor: Mostafa Ghasemi Baboli

Received: 1 October 2021
Accepted: 21 December 2021
Published: 27 December 2021

Publisher's Note: MDPI stays neutral with regard to jurisdictional claims in published maps and institutional affiliations.

Copyright: © 2021 by the author. Licensee MDPI, Basel, Switzerland. This article is an open access article distributed under the terms and conditions of the Creative Commons Attribution (CC BY) license (https://creativecommons.org/licenses/by/4.0/).

1. Introduction

Design for sustainability first started out of concern for environmental issues; it has now gone beyond moral and ideological dimensions, and has recently evolved from a general clean production method to a focus on products, and includes the social, economic, and environmental factors of production [1].

Sustainability practices aim to increase successful interdependence between human societies and ecological systems [2]. Kozlowski et al. stated that in addition to the performance in the three aspects of environmental, social, and economic sustainability, aesthetic and cultural dimensions should also be considered [3]. However, sustainable fashion products tend to be indifferent to individual taste, lifestyle, and user properties that relate to cultural environments [4,5]. For Gwilt, from a sustainable life-cycle point of view, if users' opinions and experiences were reflected in and applied to design, this could lead to a sustainable fashion development method [6].

Sumter et al. presented several sustainable design approaches, among which are the nature-inspired design and design for social innovation that this study focuses on [7]. Being inspired by and imitating nature has long been considered a valid approach to design [8], and design methods inspired by the natural environment and living culture can serve as sustainable design approaches. [9,10]. These design approaches include direct, indirect, and symbolic experiences of nature, use of natural materials, small ecological footprints, and design in relation of the ecology of place, culture, and history [8,9,11,12]. In this respect, this study conducts research on sustainable fashion products with a focus on Korean aesthetics, which is related to both sustainability and philosophy related to nature.

Oriental philosophy implies a symbiosis between nature and humans, including Taoism—which contains ideas that are consistent with nature—,Buddhism—which believes that because everything exists in relation to each other, there is no independent entity— and Confucianism—which contains the idea that nature and hu-mans are one [13,14]. This philosophy is reflected in traditional Korean houses and lifestyles, and is handed down to future generations via sustainable design that pursues symbiosis and coexistence with nature and ecological conservation. Culture and technology are inextricably linked; thus, sustainability combined with social considerations and improved design can address environmental challenges [15]. Computer technologies that facilitate innovative social transitions enable sustainable design processes with a focus on highly efficient solutions that minimize environmental impact, as covered in previous studies [7,15]. By manufacturing products at the last stage of the computer aided process plan, 3D printing enables a sustainable approach to design. The development of digital technology as a design expression tool increases user attention on and participation in the design. Digital technology expands design tools for society. For example, improvements in sustainability are achieved by encouraging interactions between consumers and designs. In this regard, 3D printing employs additive manufacturing (AM), an innovative technology that produces zero waste [16,17].

The purpose of this study is to present a sustainable fashion design approach using the image of Korea pursuing coexistence with nature as a knowledge-based database. In order to increase the emotional durability of consumers, a parametric design methodology through knowledge-based data in the 3D printing fashion product design process was used. Fashion products modeled through this process enable the production of eco-friendly products that pursue net zero through 3D printing using eco-friendly filaments.

2. Theoretical Background
2.1. Sustainability and Local Culture

In 1996, UNESCO published the report "Our Creative Diversity", which emphasizes the relationship between culture and sustainable development. The UNESCO Intergovernmental Conference on Cultural Policies for Development was held in Stockholm in 1998. Similar to the 1996 report, the correlation between culture and sustainability was emphasized throughout this conference; in particular, the significant contribution that creativity and cultural diversity make to sustainable development [18]. Moreover, "The Universal Declaration on Cultural Diversity" note that cultural diversity is as necessary to humanity as biodiversity is to nature. Cultural diversity is the root of economic growth as well as intellectual, sentimental, ethical, and mental achievements [19]. At the 2005 UNESCO Convention, the protection, promotion, and maintenance of cultural diversity were deemed essential to present and future sustainable development [20], suggesting that sustainability extends beyond environmental, economic, and social domains to include the cultural domains. There has been increasing attention among academics on the search for elements that correspond to consumer sensibility in sustainable development [18,21–23]. This study seeks to extend the domains covered by sustainability beyond environmental, social, and economic aspects to include cultural aspects.

Korean aesthetics incorporate the aesthetics that adapt to nature [24], and designs reflecting local cultures can increase emotional durability [22]. The Korean aesthetics see nature and humans as single entity, values harmonious living, and believe deeply in sharing with future generations [14]. Korean aesthetics emphasize nature-inspired beauty without embellishment by maintaining and utilizing the shape, texture, and color of nature [25]. Korean aesthetics align with the concepts of environmental preservation of sustainability, coexistence through natural circulation, the developmental evolution of ecosystems, and the pursuit of coexistence between humans and nature. For fashion products, the lifespan of the product is often determined by the empathy of and relationship with the user [6]. Objects designed to reflect local culture can build empathy and meaning for users, enabling a design that considers the user's sensibility to practice sustainability by strengthening the emotional durability of the product.

2.2. Korean Aesthetics

2.2.1. Unique Korean Folk Art

Koh, a pioneer of aesthetics in Korean art history, developed a theory on the subject of aesthetics in human life [26]. Koh claimed that Korean aesthetics reject artificial and sophisticated perfectionism and that traditional Korean art favors materials obtained from nature. He specified that Korean aesthetics have properties of folk art in that they enact understanding of cultural life sensibility. Artless art, unplanned planning, lack of refinedness, unevenness, indifference, and savory taste, which Koh considers the characteristics of traditional Korean aesthetics, mostly stem from Taoist thinking [27]. Taoism argues that the beauty of nature has achievements despite there being no artificial conduct. Moreover, rather than being guided by an artificial value system, the existence of nature itself should be the model [28]. This is similar to the study by Sumter et al. [7]. Figure 1a shows the Korean lotus design roofing tile, which Kim describes as "humane" due to its softness and mildness. Koreans admire calm and peaceful natural atmospheres without artificial decoration [29]. In Korean folk art, practical and reasonable thinking are found in unification [30]. Figure 1b shows a traditional handmade ramie *Jogakbo*, a traditional practical textile constructed by joining pieces of fabric leftover from making clothes from the days when materials were scarce. Similar to a modern recycling design, *Jogakbo* was used not only for packaging household items or for wrapping things when moving, but also for decorative and religious purposes, making it a multifunctional item. When many pieces are connected, it symbolizes longevity and has the function of emotional durability.

(a) (b)

Figure 1. Lotus design roofing tile (**a**)—source: The National Museum of Korea (https://www.museum.go.kr/site/main/relic/search/view?relicId=116730, accessed on 21 June 2021) and Ramie *Jogakbo* (**b**)—source: The National Museum of Korea (https://www.museum.go.kr/site/main/relic/search/view?relicId=205909, accessed on 21 June 2021).

2.2.2. Resilience to Nature

Some properties of Korean aesthetics include valuing creativity that conveys human touch and resilience that harmonizes with nature. Resilience in Korea means that the environment and objects naturally come together, leading to an optimistic and naturalistic attitude [31]. Such attitudes align with the Buddhist concept of holism, which claims true beauty exists in the realm of no distinction between beauty and ugliness [31].

Resilience to nature uses minimal artificial lines and planes to preserve the surrounding environment and harmoniously blend with nature. Figure 2, which depicts common and traditional folk housing called *Choga*, provides an example of this. The curve of the round roof connects with the neighboring houses, in harmony with the line of the natural environment. Resilience to nature is affected by one's subconscious and holistic thought before choosing to perceive beauty or ugliness. In Buddhist thought, this is a state where the mind is immersed in one place and forgets about the self. Here, contrasting values are extolled not to be viewed dualistically but rather as a unified, monistic whole that perceives humans as part of nature [25].

Figure 2. Korean *Choga* (photograph taken by the author).

2.2.3. Simplicity Resembling Nature

Art historian Choi emphasizes that Korean art is replete with modest, silent, elegant goodwill and noble beauty [32]. In his book "History of Korean Art", Eckardt notes that Korean art embodies simplicity above all else [33]. Eckardt recognizes the Korean aesthetics for its classic trait of having a sense of nature inspired beauty. Similarly, he recognizes its external lucidity, naturalness without exaggeration or distortion, simplicity without greed, and moderate and calmness without excess [34]. Figure 3 shows a 17–18th century jar, named the "moon jar" as it resembles the moon. The jar is simple; it is "neat", without decorations, and is a simple white color. Unnecessary elements are excluded in favor of achieving the best possible "simplicity resembling nature" [29].

Figure 3. Moon jar. Source: The National Museum of Korea (https://www.museum.go.kr/site/main/relic/search/view?relicId=941, accessed on 21 June 2021).

Figure 4 shows the oriental philosophy that influences Korean aesthetics and sustainable design through Korean aesthetics. Oriental philosophy that influences Korean aesthetics, including Taoism, Buddhism, and Confucianism, is also in contact with the concept of sustainability. This can be summarized as characteristics of Korean aesthetics including unique Korean folk art, resilience to nature, and simplicity resembling nature.

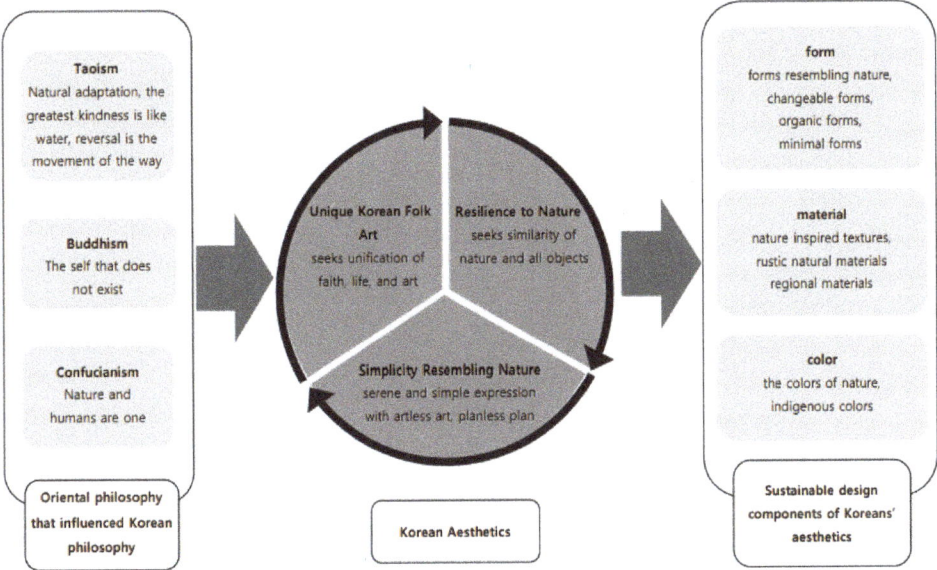

Figure 4. Oriental philosophy influencing Korean philosophy, aesthetics, and sustainable design. Source: the author.

2.3. Parametric Design

Parametric design is a computer expression method that creates a shape by parameterizing design properties into geometric elements. New shapes can be developed by changing parameters without erasing or redrawing the modeled design [35,36]. A parametric model is a computed design of geometric elements with fixed properties and other properties that can be changed. Variable properties are also called parametric properties, and fixed properties are called constraining properties. Designers change the parameters of a parametric model to find alternative solutions to the problems at hand [35]. In most CAD systems, procedural information about how a designer creates a design model is basically kept at the most basic level to implement functions such as do, undo, and redo, and few tools are provided for manipulating the process. Script editors and interpreters mostly operate as extensions and customizations of the system and are not properly designed for use by designers for modeling [37]. Therefore, it is necessary to present a parametric design process, which models fashion products. A parametric process is created with an algorithm that implements the designer's ideas. The generated algorithm can create and transform various instances by adjusting the original parameters such as the position, scale, and angle of the shape according to the concept, and the user can obtain an output suitable for the intention through simulation [38]. Yoo et al. said that a parametric design process based on the geometric principle of morphogenetic technique and the aesthetics obtained from nature can bring about this creative design result [39]. In this study, by using cloud technology to store modeled data, Fusion 360, which has the advantage of being easy to share with collaborators, was used as software.

3. Research Methods

This study aimed to develop sustainable fashion design products based on the characteristics of Korean aesthetics, including the concept of sustainability. Modeling data for 3D printing product development are calculated in the parametric design process, and it is possible to develop products with emotional durability through user intervention along with knowledge-based data of Korean aesthetics. Parametric design is widely used in software modeling. It allows various design alterations without deletion or a need to redraw [35]. Parametric modeling can directly convert the numerical values set up by the designer [40]. The designer creates an algorithm suitable to the concept by extracting the concept into a visual idea based on knowledge of art culture theory. Here, a basic gene is formed in the design. This gene communicates the generation of not only the general geometric structures but also the basic, underlying patterns of the design. In this study, a parametric design method was used, in which the designer adjusts the numerical value of the product model modeled by the designer based on the knowledge base of Korean aesthetics with the concept of sustainability.

3.1. Research Process

The sustainable product design development process is summarized in Figure 5, which was developed by the researchers using the parametric methodology concept based on Figure 5. This process explored sustainable design in Korean aesthetics and developed it into a knowledge base, with parametric components related to product size and material established. In this study, a necklace was selected as the type of fashion product. The reason is that various design changes are possible by manipulating the various pendant shapes and assembly styles that designers create through modeling data. Designers can reflect user preferences and tastes when it comes to parametric control of pendant size and assembly style.

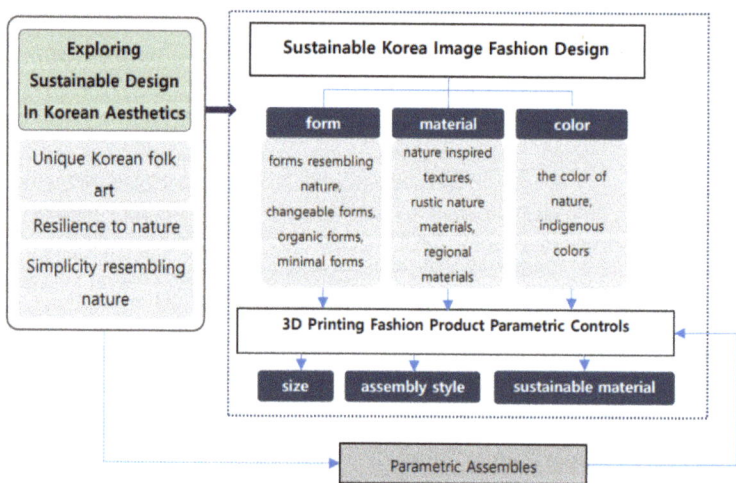

Figure 5. The 3D printing fashion product design process based on Korean aesthetics. Source: the author.

3.2. Three-Dimensional Printing Product Design Development Tools

In this study, the software program, printer, and sustainable filaments were considered as development tools for designing sustainable fashion products. Fusion 360: (2020 version, Autodesk, San Rafael, CA, USA) software was used, as it has the advantage of utilizing cloud technology that can store data and can be shared with collaborators to reduce time and carbon footprint [41]. The sustainable 3D printing filament used in this study is not a commercially available. Because it proved difficult to achieve compatibility with manufactured 3D printers, the researcher built a modular 3D printer to print the sustainable wood filament. There are many sustainable 3D printing filaments on the market, including beer filament [42], which is produced from waste by-products from the beer brewing process, recycled filament [43], based on recycled plastic, algae-based filament [44], Pāua shell filament [45], and wood filament [46]. All products printed with the wood filament are environmentally friendly in that they decompose when placed in the compost or are sent to a landfill [47]. Wood filament was selected because it had the least plastic "feel" among sustainable filaments, and the surface feeling of the print was considered to capture the simplicity of Korean aesthetics. In addition, it may stimulate the emotional durability of consumers of sustainability, who prefer a sense of "friendliness" to nature, as it exhibited less artificiality of feeling and showed the natural properties of wood.

4. Results

4.1. Three-Dimensional Printing Fashion Products Design Process Based on Korean Aesthetics

4.1.1. Exploring Sustainable Design Based on Korean Aesthetics

As discussed above, the sustainable fashion design characteristics discovered in Korean aesthetics were unique Korean folk art, resilience to nature, and simplicity resembling nature. The design of the case product was inspired by the lotus design in Figure 1. This pattern was extracted from roof tile design—an element of Korean architecture. The lotus pattern constitutes unique Korean folk art, and although it is not perfect, it expresses Korean beauty in that it is easy to compute with a symmetrical pattern and there are few artificial elements. This image was translated into digital data. Then, the parametric function and the limit scale were set up through the 3D software program.

4.1.2. Parameter Control

The components of sustainable fashion design are form, material, and color. The properties of form that reflect Korean aesthetic consciousness are "form resembling", "nature", "variable form", "organic form", and "minimal form". The material properties are "the texture of nature", "rustic nature materials", and "regional materials". The color properties are "the colors of original nature" and "indigenous colors". Based on this, size, the assembly style of the necklace, and sustainable filament were set as the moderator variable for the sustainable properties of the necklace. By limiting the associated numerical value during the fashion product modeling process, the internal design shape was configured to not invade the outline. This allows various size graftings to be initiated within the given limits, and the variables can be adjusted to reflect the user's taste in line with the program designed by the designer and can be changed into various designs.

4.1.3. Three-Dimensional Printing Process

Figure 6 shows the six steps involved in the 3D printing of the fashion product.

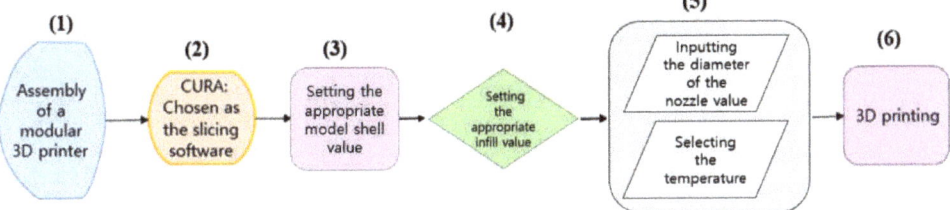

Figure 6. A concept diagram for the process of 3D printing. Source: the author.

The first step was the assembly of a modular 3D printer. Sustainable 3D printing filaments are not yet widely used, and compatibility with existing 3D printers is low. Hence, a modular 3D printer was developed. Exploration of a slicing program that can insert an optimal output value depending on the filament substance was required. The narrow nozzle of the manufactured 3D printer clogged when the wooden substance combined with PLA during printing, implying a need to develop a modular 3D printer using open-source 3D printer data. The nozzle of manufactured 3D printers is normally 0.4 mm thick. The nozzle becomes damaged when using a foreign filament, after which there is no guarantee of quality. This study used a 0.6 mm nozzle instead. The newly assembled printer required tuning of the output data for normal operation. In the second step, the slicing software was explored. To 3D print the final digital data, CURA: (version 15.04.6, Ultimaker, Utrecht, The Netherlands) was chosen, as it ensures that the form has strong durability even with a minimal amount of 3D data. CURA also helps the 3D printer use the proper number of filaments without waste. The third step consists of setting the appropriate model shell value for product stabilization, post-printing. In the fourth step, the appropriate infill value was set to maintain a stable form with minimal filament consumption. The fifth step involved inputting the diameter of the nozzle value and selecting a temperature suitable for the filament type. To prevent contraction caused by a temperature difference during the printing process, the bed temperature was heightened. In the sixth step, the researchers printed from the 3D printer using the sustainable filament.

In a previous study, the traditional texture was reproduced through wood, stone, soil, and more. That study utilized local and endemic materials that constitute sustainable Korean materials. In the case of wood filaments, the wooden substance is already ground down during wood processing; this shortens the filament-production process and allows for recycling. Filaments containing wood and PLA show high biodegradability and are thus eco-friendly. A major reason to use the wood filament is the outstanding aesthetic of the final products in comparison to plastic products. The wood filament in this study produced natural wood grain-like texture, and the color required no post-processing. The 3D printing filament printing process is shown in Figure 7.

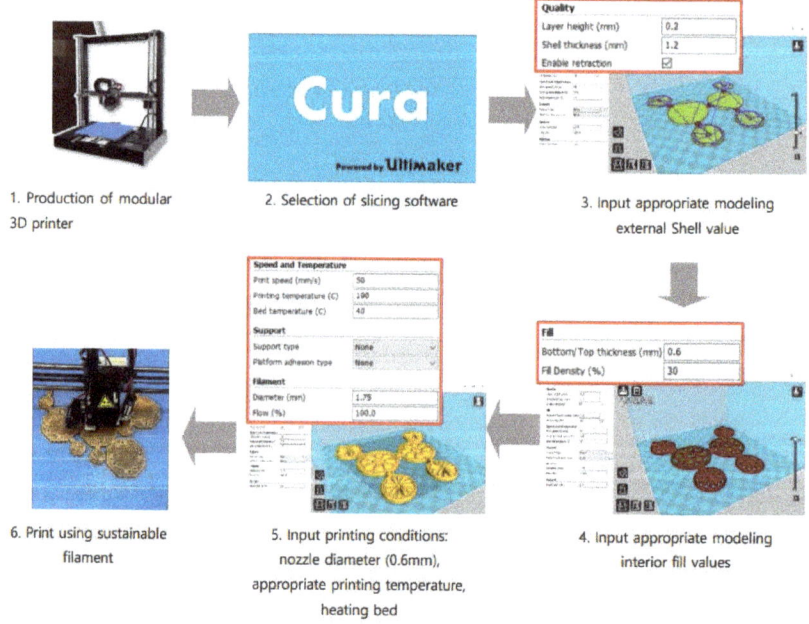

Figure 7. The 3D printing process. Source: the author.

4.2. Three-Dimensional Printing Necklace Based on Korean Aesthetics Development Results

In this study, a necklace with emotional durability was developed for those who prefer the Korean philosophy of coexistence with nature and nature-inspired design. To this end, the pendant of the necklace was designed with the image of a roofing tile with a lotus flower pattern; this is easy to apply the parametric design methodology to, is based on original and unique Korean folk art, and has an assembly for which the design can be changed in line with the user's choice. In Korea, the lotus flower symbolizes Buddhism and has the meaning of always being clean, even in a dirty place [48]. This is similar to Korea's resilience to nature aesthetics, which finds beauty by adapting to nature. The lotus pattern is used not only in Korean architecture, but also in stationery, crafts, and folk crafts such as ceramics [49]. In addition, it may strengthen a user's emotional durability as it is often used as a pattern to pray for family happiness [50].

In this study, a necklace design was proposed with two roofing tile images with different lotus patterns, shown as "1" and "2" in the first column of Figure 8. Model creation was initiated by digitizing the design sketch. A total of two pendant designs were developed from these image data. The 3D print digital data design parametric components were shape and size, assembly style, and combination. The two pendant designs are shown as "3" and "4" in the shape column of the 3D printed digital data design parameter control in Figure 8. The parametric function was set in the digitalized model, and the limit range was set depending on the function. This setting and range limit allowed automatic adjustment of the connected flower's petal and anther size by adjusting the size of the outline. Parametric variable adjustment of the desired area can change the size of each part. I proposed 3-1, 3-2, 4-1, and 4-2 pendants in the size section by changing the dimensions of pendants 3 and 4. In this study, only four samples were made in the size section, but more size conversions are possible according to the designer's intentions. There are two types of assembly styles for connecting pendants, suggested as "5" and "6" of the assembly column. The design can be changed according to the user's taste by suggesting different types of assemblies as modeling data. For the connecting parts, the 3D printer's assembly function was used, with Fusion 360 selected as the 3D digital software. Fashion products require an assembly process but when the assembly function is used, it can be planned in the digital data. This allows assembly simulation before printing. Therefore, it is possible to print all at once in the assembled state. The assembly function allows a sustainable production process by saving raw material use, allowing for zero waste, and reducing production time and cost. The assembly design was designed into hinge and chain style. If the assembly style connecting the pendant is designed in more styles, the diversity of designs can be increased.

In this study, six necklace designs were proposed—ND.1, ND.2, ND.3, ND.4, ND.5, and ND.6, as shown in the combination column. Depending on how the pendant and assembly style are combined, various design changes are possible. For example, ND.1 is a necklace design made using 3-1 and 3-2 pendants and assembly style 5. ND.2 was completed by combining size-adjusted 3-2 and 4-1 pendants with assembly style 6. ND.3 used size-adjusted 3-1, 3-2, and 4-1 pendants and assembly style 5. Connecting multiple pieces or reducing the number of pieces stimulates the user's emotional durability.

Figure 8 depicts the design process, showing the development process for a 3D printed necklace based on Korean aesthetics, through a parametric design methodology and process.

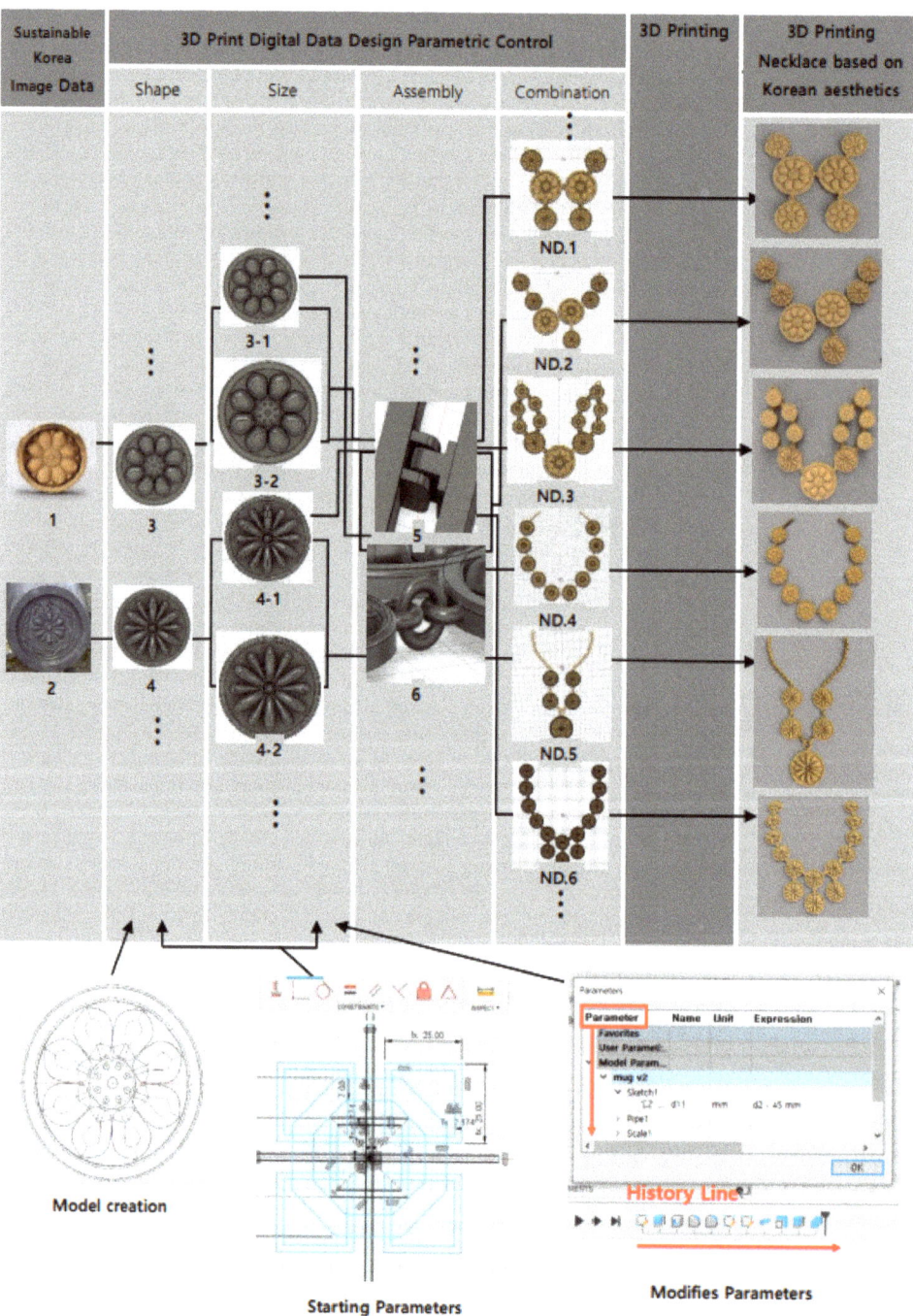

Figure 8. Three-dimensional printing of necklaces based on Korean aesthetics. Source: the author.

5. Conclusions

Various attempts are being made to develop a sustainable design process that changes with the development of digital technology. While the popularity of sustainability is affecting the fashion industry, users are dissatisfied with the level of sustainable design currently available. Therefore, this study proposed an objective process by creating an algorithm that increases the emotional attachment of users by incorporating a sustainable Korean aesthetic into the design process created by the designer's intuition. To find a creative expression method for sustainable design, the direction and relevance of sustainability were explored via Korean aesthetics. With an adaptive thinking that considers nature and humans together, the Korean aesthetic sense of living in harmony with the environment exhibits many elements of current trends in sustainable development. This has had an impact on Korean life in general, and has helped to solve the problem of creative elements of Korean design.

This study is meaningful in that it formulates an objective process by conceptualizing parts that could not be objectified only by design sensibility, extracting them as visual ideas based on knowledge of art and culture theory, and creating algorithms suitable for concepts. This process allows users to participate in the production process, thereby satisfying their sensibility and taste, and hence incorporating the aesthetic and cultural dimensions of sustainability, which were previously neglected. Through this process, the user selects a preferred variable among those set by the designer, thereby increasing the degree of attachment to the selected product by increasing participation compared with existing products.

This study is also meaningful in that it extends the product life-cycle through this attachment. The results of this study, which digitized Korean images against the backdrop of Korean aesthetics with flexible adaptability while coexisting with nature, were able to develop various sizes and designs via parametric numerical control, which is expected to expand the scope of emotional durable product development by providing users with limited data. In addition, it is expected that implementing assembly functions with digital data will reduce the time and cost required for the manufacturing process and help practice sustainable fashion products in that it attempted to modularize products. However, for the results of this study to become practical, a follow-up study is needed to investigate the users of this product.

Funding: This research received no external funding.

Data Availability Statement: Data available on request due to restrictions e.g., privacy or ethical.

Conflicts of Interest: The author declares no conflict of interest.

References

1. Clark, G.; Kosoris, J.; Hong, L.N.; Crul, M.R.M. Design for sustainability: Current trends in sustainable product design development. *Sustainability* **2009**, *1*, 409–424. [CrossRef]
2. Spangenberg, J.H.; Fuad-Luke, A.; Blincoe, K. Design for Sustainability (DfS): The interface of sustainable production and consumption. *J. Clean. Prod.* **2010**, *18*, 1485–1493. [CrossRef]
3. Kozlowski, A.; Bardecki, M.; Searcy, C. Tools for Sustainable Fashion Design: An Analysis of Their Fitness for Purpose. *Sustainability* **2019**, *11*, 3581. [CrossRef]
4. Agost, M.-J.; Vergara, M. Principles of Affective Design in Consumers' Response to Sustainability Design Strategies. *Sustainability* **2020**, *12*, 10573. [CrossRef]
5. Rahman, O.; Gong, M. Sustainable practices and transformable fashion design—Chinese professional and consumer perspectives. *Int. J. Fash. Des. Technol. Educ.* **2016**, *9*, 233–247. [CrossRef]
6. Gwilt, A. Valuing the Role of the Wearer in the Creation of Sustainable Fashion. *Res. J. Text. Appar.* **2013**, *17*, 78–86. [CrossRef]
7. Sumter, D.; De Koning, J.; Bakker, C.; Balkenende, R. Circular Economy Competencies for Design. *Sustainability* **2020**, *12*, 1561. [CrossRef]
8. Ceschin, F.; Gaziulusoy, I. Evolution of design for sustainability: From product design to design for system innovations and transitions. *Des. Stud.* **2016**, *47*, 118–163. [CrossRef]
9. Dias, B.D. Beyond sustainability–biophilic and regenerative design in architecture. *Eur. Sci. J.* **2015**, *11*, 147–158.

10. Ramzy, N. Sustainable Spaces with Psychological Values: Historical Architecture as Reference Book for Biomimetic Models with Biophilic Qualities. *Archnet-IJAR Int. J. Arch. Res.* **2015**, *9*, 248. [CrossRef]
11. Naderi, J.R.; Kellert, S.R.; Heerwagen, J.H.; Mador, M.L. *Biophilic Design: The Theory, Science and Practice of Bringing Buildings to Life*; John Wiley and Sons: Hoboken, NJ, USA, 2009; Volume 93, pp. 262–265.
12. Joye, Y. Cognitive and Evolutionary Speculations for Biomorphic Architecture. *Leonardo* **2006**, *39*, 145–152. [CrossRef]
13. Lee, S.J. Study on Monistic Design Thought and Method. *Arch. Des. Res.* **2004**, *17*, 479–486.
14. Yang, E.J.; Kim, K.H. The Study on Characteristics of Korean Traditional Space from an Ecological and Aesthetic Point for View. *Korean Inst. Inter. Des. J.* **2010**, *19*, 57–66.
15. Ramani, K.; Ramanujan, D.; Bernstein, W.Z.; Zhao, F.; Sutherland, J.; Handwerker, C.; Choi, J.-K.; Kim, H.; Thurston, D. Integrated Sustainable Life Cycle Design: A Review. *J. Mech. Des.* **2010**, *132*, 091004. [CrossRef]
16. Gebler, M.; Uiterkamp, A.J.S.; Visser, C. A global sustainability perspective on 3D printing technologies. *Energy Policy* **2014**, *74*, 158–167. [CrossRef]
17. Atzeni, E.; Salmi, A. Economics of additive manufacturing for end-usable metal parts. *Int. J. Adv. Manuf. Technol.* **2012**, *62*, 1147–1155. [CrossRef]
18. Soini, K.; Birkeland, I. Exploring the scientific discourse on cultural sustainability. *Geoforum* **2014**, *51*, 213–223. [CrossRef]
19. The United Nations Educational, Scientific and Cultural Organization (UNESCO). UNESCO Universal Declaration on Cultural Diversity 2 November 2001. In *Standard-Setting at UNESCO*; The United Nations Educational, Scientific and Cultural Organization (UNESCO): Paris, France, 2007; pp. 707–712.
20. The United Nations Educational, Scientific and Cultural Organization (UNESCO). Convention on the Protection and Promotion of the Diversity of Cultural Expressions 2005. In Proceedings of the Paper Presented at the General Conference of the United Nations Educational, Scientific and Cultural Organization, Paris, France, 3–21 October 2005.
21. Burcikova, M. One Dress: Shaping Fashion Futures through Utopian Thinking. *Fash. Pr.* **2019**, *11*, 328–345. [CrossRef]
22. Sandhu, A. Fashioning Wellbeing Through Craft: A Case Study of Aneeth Arora's Strategies for Sustainable Fashion and Decolonizing Design. *Fash. Pract.* **2020**, *12*, 172–192. [CrossRef]
23. Armstrong, C.M.; Niinimäki, K.; Lang, C. Towards Design Recipes to Curb the Clothing Carbohydrate Binge. *Des. J.* **2016**, *19*, 159–181. [CrossRef]
24. Fletcher, K. Durability, Fashion, Sustainability: The Processes and Practices of Use. *Fash. Pract.* **2012**, *4*, 221–238. [CrossRef]
25. Choi, M. A Study on the 'Win-Win' Design Direction for Nature and Human Beings with Special Reference to Oriental Ideas and Korean Aesthetic Beauty. *Bull. Korean Soc. Basic Des. Art* **2010**, *11*, 477–490.
26. Korean Association of Industrial Designers. *Korean Design Application Case Study*; Ministry of Trade, Industry and Energy: Gwacheon, Korea, 1997.
27. Cho, M. Philosophy: Confucianist and Taoist Aesthetic Contemplation on the Study of the Features of Korean Aesthetic Beauty. *Korean Philos. Cult.* **2012**, *65*, 299.
28. Kim, S.H.; Hur, B.P. A Study on Sustainable Earth Architecture Characteristics from an Ecological Aesthetic Point of View— Around the Thoughts of Lao-tzu. *Korean Inst. Inter. Des. J.* **2011**, *20*, 54–62.
29. Choi, J. *Korean Beauty, the Aesthetics of Freedom: Why Korean Beauty Here and Now*; Hyohyung: Seoul, Korea, 2000.
30. Park, J. Chosun's Crafts and Yanagi Muneyoshi's Viewpoint of Tea-Tao. *J. Tea Cult. Ind. Stud.* **2006**, *3*, 1–13.
31. Kim, W. *Exploring Korean Beauty*, 6th ed.; Youlhwadang: Seoul, Korea, 1996.
32. Kwon, Y. *Re-reading Korean Beauty: 100 Years Discussion on Korean Beauty Seen through 12 Aestheticians*, 6th ed.; Dolbegae: Paju, Korea, 2005.
33. Eckardt, A. *A History of Korean Art*; Hiersemann: Leipzig, Germany, 1982.
34. Kwon, Y. Andreas Eckardt's Art History Perspectives. *Korean Bull. Art Hist.* **1992**, *5*, 5–31.
35. Hernandez, C.R.B. Thinking parametric design: Introducing parametric Gaudi. *Des. Stud.* **2006**, *27*, 309–324. [CrossRef]
36. Lotfabadi, P.; Alibaba, H.Z.; Arfaei, A. Sustainability; as a combination of parametric patterns and bionic strategies. *Renew. Sustain. Energy Rev.* **2016**, *57*, 1337–1346. [CrossRef]
37. Shih, S.-G. Notes on Generative Modeling, Procedural Symmetry, and Constructability of Architectural Design. *Comput. Des. Appl.* **2014**, *11*, 518–525. [CrossRef]
38. Abdullah, H.K.; Kamara, J.M. Parametric Design Procedures: A New Approach to Generative-Form in the Conceptual Design Phase. *Archit. Eng. Conf.* **2013**, *201*, 334–343.
39. Yoo, Y.S.; Cho, M.; Eum, J.S.; Kam, S.J. Biomorphic Clothing Sculpture Interface as an Emotional Communication Space. *Front. Psychol.* **2020**, *11*, 117. [CrossRef] [PubMed]
40. Verroust, A.; Schonek, F.; Roller, D. Rule-oriented method for parameterized computer-aided design. *Comput. Des.* **1992**, *24*, 531–540. [CrossRef]
41. Total 3D Printing. Fusion 360 vs AutoCAD [2021]: Which CAD Software Is Best? Available online: https://total3dprinting.org/fusion-360-vs-autocad-compared/ (accessed on 21 June 2021).
42. 3D Print with Beer Filament [Internet]. 2015. Available online: https://all3dp.com/3d-print-with-beer-filament/ (accessed on 21 June 2021).
43. Pinshape. Guide to Green 3D Printing—4 Ways to Be More Sustainable! Available online: https://pinshape.com/blog/guide-green-3d-printing/ (accessed on 21 June 2021).

44. Global Opportunity Explorer. Algae-Based 3D-Printing Filaments. Available online: https://goexplorer.org/algae-based-3d-printing-filaments/ (accessed on 21 June 2021).
45. Scion Research (YouTube). Scion Paua Power. Available online: https://youtu.be/lvxshNCNFp4 (accessed on 24 August 2021).
46. 3DINSIDER. Wood Filament: Properties, How to Use Them, and Best Brands. Available online: https://3dinsider.com/wood-filament/ (accessed on 21 June 2021).
47. 3D Printer and 3D Printing News. Jelwek Launches 3D Printed, Wood Filament Watch Collection. Available online: https://www.3ders.org/articles/20141214-jelwek-launches-3d-printed-wood-filament-watch-collection.html (accessed on 21 June 2021).
48. Kim, J.; Yoon, P.; Kim, Y. Geometric and Design Form, Container Type, and Plant Materials of the Korean Traditional Buddhist Flower Arts. *J. Korean Soc. Flor. Art Des.* **2008**, *18*, 7–29.
49. Kim, Y.H. Phenomenological Meaning of Lotus Flower Pattern Symbolism. *Korea Sci. ART Forum* **2013**, *12*, 25. [CrossRef]
50. Jang, M.; Hong, J. The study of Wedding Dress by Applying the Pattern of Traditional Wedding Dress—Focused Lotus flower, Arabesque, Peony Patterns. *J. Korean Soc. Fash. Beauty* **2008**, *6*, 97–103.

 sustainability

Article

Chemical Recycling of a Textile Blend from Polyester and Viscose, Part II: Mechanism and Reactivity during Alkaline Hydrolysis of Textile Polyester

Jenny Bengtsson [1,*], Anna Peterson [1], Alexander Idström [2], Hanna de la Motte [1] and Kerstin Jedvert [1,*]

1 RISE Research Institutes of Sweden, Argongatan 30, SE-431 22 Mölndal, Sweden
2 Department of Chemistry and Chemical Engineering, Chalmers University of Technology, SE-412 96 Gothenburg, Sweden
* Correspondence: jenny.bengtsson@ri.se (J.B.); kerstin.jedvert@ri.se (K.J.)

Abstract: Chemical recycling of textiles holds the potential to yield materials of equal quality and value as products from virgin feedstock. Selective depolymerization of textile polyester (PET) from regenerated cellulose/PET blends, by means of alkaline hydrolysis, renders the monomers of PET while cellulose remains in fiber form. Here, we present the mechanism and reactivity of textile PET during alkaline hydrolysis. Part I of this article series focuses on the cellulose part and a possible industrialization of such a process. The kinetics and reaction mechanism for alkaline hydrolysis of polyester packaging materials or virgin bulk polyester are well described in the scientific literature; however, information on depolymerization of PET from textiles is sparse. We find that the reaction rate of hydrolysis is not affected by disintegrating the fabric to increase its surface area. We ascribe this to the yarn structure, where texturing and a low density assures a high accessibility even without disintegration. The reaction, similar to bulk polyester, is shown to be surface specific and proceeds via endwise peeling. Finally, we show that the reaction product terephthalic acid is pure and obtained in high yields.

Keywords: textile recycling; polyester; alkaline hydrolysis; depolymerization; peeling reaction

Citation: Bengtsson, J.; Peterson, A.; Idström, A.; de la Motte, H.; Jedvert, K. Chemical Recycling of a Textile Blend from Polyester and Viscose, Part II: Mechanism and Reactivity during Alkaline Hydrolysis of Textile Polyester. *Sustainability* **2022**, *14*, 6911. https://doi.org/10.3390/su14116911

Academic Editor: Joshua M. Pearce

Received: 22 April 2022
Accepted: 1 June 2022
Published: 6 June 2022

Publisher's Note: MDPI stays neutral with regard to jurisdictional claims in published maps and institutional affiliations.

Copyright: © 2022 by the authors. Licensee MDPI, Basel, Switzerland. This article is an open access article distributed under the terms and conditions of the Creative Commons Attribution (CC BY) license (https://creativecommons.org/licenses/by/4.0/).

1. Introduction

The global consumption of textile fibers is increasing annually, with polyester (PET) accounting for more than half of the total fiber volume [1]. With increased consumption follows increased waste generation, and currently textile waste is predominately sent to landfill or incineration. Globally, less than 1% of all textile waste is recycled into fibers for use in new textile products [2]. However, there are often clear environmental benefits, e.g., reduced carbon footprint and reduced energy consumption, to reuse or recycle textiles [3].

An attractive recycling route for PET waste textiles is chemical recycling through depolymerization. Depolymerization of PET textile waste produces monomers, which may substitute the use of monomers synthesized from fossil resources. PET produced by polymerization of monomers from such chemical recycling has the potential to hold an equal quality to virgin materials. The polyester derived monomers can also substitute fossil-derived chemicals in the production of other value-added materials, such as alkyd resins [4], reactive polyester resins [5], or polyurethanes [6]. This is to be compared with thermomechanical recycling, in which PET is melted and reshaped, which typically yields materials with lowered molecular weight and inferior mechanical properties [7]. Chemical recycling has also proven feasible for the separation of PET from blended textiles. Selective depolymerization of PET from cellulose/PET blends has been reported as an attractive opportunity to separate and recover the blend components for further processing into value-added products [8,9].

Depolymerization of polyester yields the monomers of the textile. The majority of polyester textiles are composed of poly(ethylene terephthalate) (PET), formed by a condensation reaction between terephthalic acid (TPA) and ethylene glycol (EG). The textile nomenclature, however, does not specify that polyester textile must be composed of PET, but also allows other types of polyesters. Further, co-monomers might be present in the polymer chain to infer certain properties, such as increased dyeability [10] or flame retardance [11]. The scientific literature available on depolymerization of textile PET is sparse, while there is a large body of literature on PET packaging depolymerization, predominantly from bottle waste [12,13]. Textile recycling is heavily underdeveloped and new solutions are urgently needed, especially considering that the volumes of polyester produced for textile purposes are twice as high as PET produced for packaging purposes. About 55 million tonnes of PET staple fibers were put on the market 2019 [14], while PET for packaging amounted to about 25 million tonnes [15]. Studies on recycling of textile PET are, hence, highly relevant.

In chemical recycling, the ester bond of polyesters is commonly cleaved by solvolytic methods; the main ones being glycolysis, methanolysis, and hydrolysis. These solvolytic methods all render small-molecule building blocks of PET with slightly different chemistries. Recently, novel methods, including enzymatic degradation of PET [16] and solid-state reaction methods [17], have been proposed; however, these are still at an early research state.

Industrially, glycolysis is the oldest and most common method for PET depolymerization [18]. It should be noted, however, that its widespread use is limited by the low profitability of PET glycolysis, or any solvolytic recycling method, due to competition with cheap virgin petrochemical feedstock [19]. Glycolysis of PET in EG produces bis(hydroxyethyl terephthalate) (BHET), which is an intermediate in the PET polymerization reaction. While it is possible to polymerize PET from BHET, it is not commonly done and manufacturers favor TPA as a feedstock for PET production due to its greater process performance [19]. The yields of the hydrolysis reaction are typically high, approaching 100% while both glycolysis and methanolysis typically have lower yields [18,20]. Furthermore, hydrolysis can be conducted under mild conditions, renders monomers of high purity, and can tolerate contaminated waste streams.

The present research covers depolymerization of textile PET using alkaline hydrolysis. Alkaline hydrolysis of textile cellulose/PET blends has the advantage that the aqueous alkaline environment causes hydrolytic cleavage of the ester bond of PET, while cellulose is relatively resistant to alkali degradation [21]. Hence, cellulose can be recovered in its polymeric form, suitable for recycling into regenerated fibers or microcrystalline cellulose, a topic discussed in Part I of this publication series. While several studies exist on the reaction conditions, kinetics and mechanism of PET depolymerization using alkaline hydrolysis [18,22–25] textile PET is neglected. Studies typically use granulates of virgin PET or post-consumer PET in the form of flakes or powders as the polyester source. Processing of PET into fibers, however, renders an increased crystallinity and a different morphology compared to bulk PET [26]. Further, PET used for textile applications has a lower molecular weight compared to PET used for bottles [27].

From the literature on nontextile polyester, there are indications that hydrolysis takes place at the external surface of the solid-state PET and that the major decomposition reaction occurs at the ends of the polymer chains [25]. López-Fonseca et al. found that the residual particle size of PET-particles decreased during hydrolysis while no structural differences were observed [22,28]. Similarly, Mishra et al. found that the molecular-weight distribution (MWD) of PET was not changed over hydrolysis times of 90 min at temperatures ranging from 25 to 150 °C [29]. A large impact of particle size on the rate of hydrolysis, with a decrease in hydrolysis efficacy as a function of increased particle size, has been reported. The rection rate of hydrolysis has been shown to be proportional to the particle surface area both for alkaline [28,29] and acidic hydrolysis [30,31].

In the present study, hydrolysis of a PET fabric is studied under mild conditions: 90 °C, 5 wt% NaOH, and atmospheric pressure. While this publication deals exclusively with the depolymerization of textile PET, the reaction conditions are optimized for separation and recovery of a cellulose/PET blend. The hydrolysis rate is explored as a function of PET accessible area, while comparing laundered and non-laundered fabric, as well as shredded fabric. The reaction products, as well as residual PET, is thoroughly characterized, using both spectroscopic and chromatographic methods.

2. Materials and Methods

2.1. Materials

The never-used PET tricot fabric was supplied by a Swedish fashion company. The fabric was either used as received or laundered, and subsequently cut into approximately 1×1 cm pieces. Washing was performed at 60 °C, according to Swedish standard SS-EN ISO 6330:2012. A portion of the washed fabric was also shredded (New Shunxing, NSX-QT 310). Sodium hydroxide (NaOH, 50%) and sulfuric acid (H_2SO_4, reagent grade, 95–97%) were obtained from Sigma-Aldrich, and acetic acid (glacial, 99%,) was obtained from Fisher Scientific and used as received.

2.2. Hydrolysis of PET

Alkaline hydrolysis of PET was performed in a 500 mL glass reactor. The NaOH concentration was 5 wt% and the weight ratio of sample to NaOH solution was kept at 1:100. The aqueous NaOH was heated to 90 °C before adding 5.0 g of oven dry (2 h, 105 °C) sample to the reaction vessel. Hydrolysis was performed for the selected time (60–1440 min). The reaction was quenched by immersing the reactor in an ice bath. Post reaction, the solid residue was separated from the reaction solution by filtration, using a tight-knit PET wire fabric as the barrier. The solids were neutralized in 5% acetic acid and thereafter washed with water and dried at 105 °C for 4 h to calculate the gravimetric yield. The liquid phase was kept in the freezer for further analysis.

2.3. Characterization of Fabric

All analyses on the fabric were made on the laundered sample. The fabric was first dried at 50 °C for 2 h and thereafter conditioned for at least 16 h at 25 °C, 65% relative humidity.

The density (GSM) was determined from 100 cm^2 samples punched out from different sampling areas on the fabric. Samples were weighed and the density (g m^{-2}) was calculated. A total of 5 sampling points were used.

The thickness of the fabric was measured according to SS-EN ISO 5084, using a Shirley thickness tester. The sample area was 100 cm^2, and the pressure was 1 kPa. A total of 5 sampling points were used.

The structure of yarns and fibers were studied using a microscope (Nikon Eclipse). Prior to examination the fabric was disintegrated by means of carding to reveal the individual yarns and fibers.

2.4. Characterization of Solid Residue

Molecular-weight determination was performed by Smithers, UK. A total of 20 mg of solid residue was dissolved in hexafluoropropan-2-ol overnight. The solution was then filtered through a 0.45 μm PTFE membrane into autosampler vials and inserted into the column (Malvern/Viscotek TDA 301) equipped with a refractive index detector. Poly(methyl methacrylate) was used for calibration.

Wide-angle X-ray scattering (WAXS) measurement was performed at Chalmers University of Technology (Mat:Nordic, SAXSLAB) with a 0.9 mm beam diameter Rigaku 003+ high brilliance microfocus Cu-radiation source at 130 mm distance with the sample irradiated over 20 min.

2.5. Characterization of Filtrate

The TPA-concentration in the filtrate was measured using UV-absorption at 242 nm by UV spectrophotometry (UV spectrometer, Analytik Jena). A linear calibration curve was constructed from 5 measurements of 2 to 10‰ TPA in 5% NaOH.

Further, a portion of each filtrate was carefully weighed and acidified to pH 2–3 by the addition of 4 M H_2SO_4, which causes TPA to precipitate. The TPA precipitates were separated using a glass microfiber filter and rinsed with distilled water, and then the weight of TPA was determined after drying for 4 h at 105 °C.

To study the degradation products, nuclear magnetic resonance (NMR) spectroscopy was used. A small-scale hydrolysis of PET in 5% NaOH and D_2O was performed to reduce the water signal obtained in the spectra. The reaction conditions were, in other aspects, identical to the large-scale experiments, i.e., 90 °C, 5 wt% NaOH and reaction times between 60 and 1440 min. The hydrolysis filtrate was analyzed with NMR spectroscopy. The NMR measurements were conducted on a Varian 400-MR spectrometer operating at 9.4 T, equipped with a OneNMRProbe. The ^{1}H-NMR spectrum parameters included a 5 ms ^{1}H-detection pulse, 2.5 s acquisition time, 2 s recycle delay, and 32 scans. The samples were studied using D_2O as solvent, and the chemical shifts were referenced to the residual solvent signal.

3. Results and Discussion

Three textile samples, with different accessibility, were prepared from the same virgin, never-laundered PET tricot fabric and subjected to alkaline hydrolysis (Figure 1). Sample A was made by cutting the never-laundered tricot into 1 × 1 cm squares. Sample B was made by laundering the sample at 60 °C to remove surface finishing and other processing aids remnant on the fabric. The sample was subsequently sectioned in the same way as sample A. Sample C was made by shredding the laundered fabric in an industrial textile shredder. During this process, the fabric is disintegrated, and the individual threads and fibers liberated, increasing the surface area. As such, samples A–C with an increasing accessibility were produced. Further, an attempt was made to grind the fabric into even smaller fragments (powder) using a rotor mill. However, this sample was not included in the hydrolysis series as, apart from the desired fine fibril grinds, part of the sample melted and formed small crumbles. The sample was excluded as this two-phase material would not be representative.

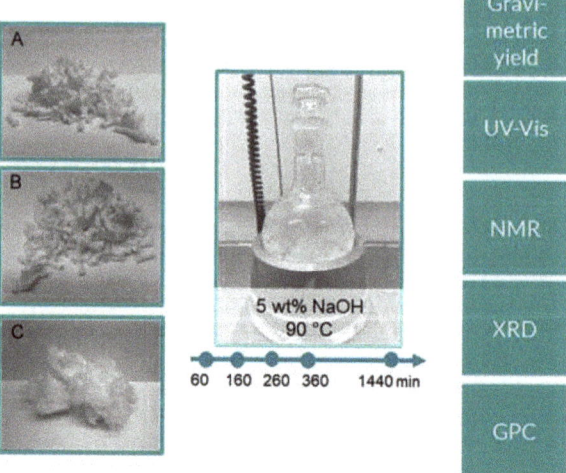

Figure 1. Schematic of the samples, reaction conditions, and characterization methods used. Sample (**A**), never-laundered, cut; sample (**B**), laundered and cut; sample (**C**), laundered and shredded.

Alkaline hydrolysis was performed under mild conditions during 60, 160, 260, 360, or 1440 min. As expected, the mass loss of samples increases with hydrolysis time (Figure 2a). After 1440 min reaction time, all samples show 100% mass loss. For sample B, an additional datapoint was taken at 960 min, which also shows 100% mass loss. We assumed an increasing accessibility along samples A–C, due to removal of finishing agents and an increase in surface area. However, the differences in mass loss between the samples at each of the measuring points is small, which was surprising.

Figure 2. (a) Mass loss of samples A, B, and C, with increasing accessibility during hydrolysis, (b) linear fits to the first 4 datapoints of each sample shown in (a).

The initial hydrolysis rate of the never-laundered sample is slightly lower compared to the laundered samples, which is signified by the flatter slope of the linear fit of sample A as compared to sample B and C (Figure 2b). This is expected as it is known that surface finishing agents or processing aids remnant on the fabric surface often decrease the rate of hydrolysis [32]. However, the accessibility of fibers in terms of surface area does not seem to influence the hydrolysis rate. This indicates that mechanical pretreatment of the PET textile to increase the surface area prior to alkaline hydrolysis does not yield a higher rate of hydrolysis. These results were somewhat surprising, as hydrolysis of PET is considered a surface specific reaction with a strong correlation between particle size and reaction rate.

We went on to study the yarn and fabric structure to explain the negligible impact of shredding on the reaction rate. The unraveled yarn was studied in a light microscope and assigned as a non-twisted, texturized multifilament. Texturing is typically performed on synthetic filaments and refers to the formation of crimp, loops, coils, or crinkles in the filament. This treatment decreases the density of yarns spun from texturized filaments and creates air pockets, which gives a different handle of fabrics, as well as improved ventilation and absorbency. The thickness of the sample at a pressure of 1 kPa was 0.6 ± 0.02 mm while the density was 132 ± 2 g m^2, which is in the lower range of knit fabrics. Given the texturing of the yarn and the low density of the fabric, the accessibility of the textile to hydrolysis should not be considerably impacted by shredding, which is in line with our findings. Texturing of synthetic yarns is a common practice throughout the fashion industry, hence, shredding of PET fabrics can tentatively be omitted without impacting the kinetics of the hydrolysis reaction. This is advantageous for the recycling scheme of textiles as it reduces the number of pretreatment steps and consequently the energy consumption.

Next, the yield of TPA was compared with the mass loss of PET, in order to determine whether PET is completely depolymerized into TPA and EG, or also yields other degradation products. The theoretical yield was calculated based on the mass loss at each time interval, assuming that PET is formed by the esterification of equimolar amounts of TPA and EG. The calculated theoretical yield is compared to the experimental yield of TPA obtained via gravimetric or spectroscopic means (c.f. Experimental). Gravimetric yields were obtained by precipitation of TPA from carefully weighed portions of the reaction filtrate. Spectroscopic yields were obtained by UV-Vis spectroscopy of the reaction filtrate at 242 nm, a wavelength where the anionic form of TPA, present under alkaline conditions, shows a strong UV absorbance.

There is an excellent agreement between the theoretical yield and the yield obtained from UV absorption spectroscopy, indicating complete degradation of PET into TPA and EG (Figure 3). Gravimetric data on the mass of precipitated TPA from the reaction filtrate also show good agreement with the theoretical yields. The conversion from degraded PET to TPA at each timepoint, hence, is close to 100%.

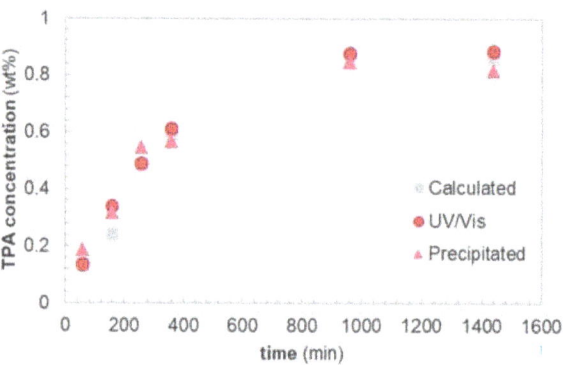

Figure 3. TPA concentration in reaction solution after depolymerization of sample B, laundered and cut.

Nuclear magnetic resonance (NMR) spectroscopy was used to further verify the reaction course and purity of products. In Figure 4 ^{1}H-NMR spectra of samples obtained after 260, 360, and 1440 min are shown.

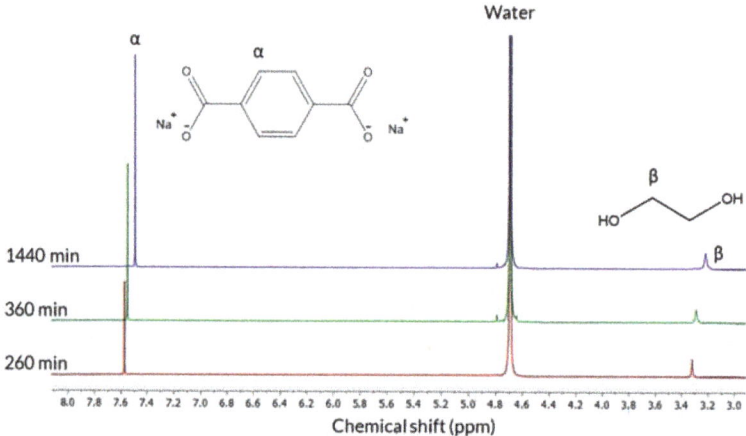

Figure 4. ^{1}H-NMR spectra of reaction solution at time points 260, 360, and 1440 min of hydrolysis in alkaline D_2O. Proton signals significative for TPA and EG are marked α and β, respectively.

For all spectra, in addition to the residual solvent signal, two singlets can be seen, one around 7.5 ppm and one around 3.3 ppm, assigned to originate from TPA and EG, respectively. An expected increase in the integrals for both components can be seen with increased hydrolysis time. For each time point, the integral of the signals from the protons complies well with each other indicating a 1:1 molar ratio of the two components, as expected from the starting material.

The presence of two singlets is distinct proof that only monomers of TPA and EG are present in the sample. If oligomers were present, a much more complicated splitting structure of the signals from the TPA and EG protons would be seen. One could imagine an oligomer where one, or the other, signal would appear as a singlet; however, the other cannot in that case appear as a singlet. One could also imagine that for a large polymer the protons will appear as two singlets; however, in addition to the fact that such a polymer would not be soluble, the narrow line width of the singlets indicates a small molecule, such as a monomer, rather than a large polymer. The end groups would, in that case, also be visible in the spectrum, which is not the case. The slight change in the chemical shift of the two components between different time points could be attributed to a change in pH, as expected upon consumption of hydroxide ions during depolymerization.

Once the effectiveness of alkaline hydrolysis on textile PET and the impact of accessibility was established, we moved on to study the degradation mechanism. TPA yields were close to the theoretical maximum at each time point during hydrolysis, indicating that the polymer is depolymerized from the chain ends via a peeling reaction, in agreement with earlier results from Kumar and Guria [25]. Further, NMR analysis showed pure TPA, with no visible traces of oligomers or other degradation products, which further supports a peeling reaction from the chain ends.

Molecular-weight distributions (MWD) were obtained by GPC for the solid residue of sample B after hydrolysis for different times. The MWD is identical for samples run for 60, 260, or 360 min (Figure 5). GPC of the sample run for 1440 min could not be performed as no solid residue was left. The overlapping curves show no changes to the MWD over the course of reaction. This indicates that chain scission, which would drastically decrease the molecular weight, is not prevalent.

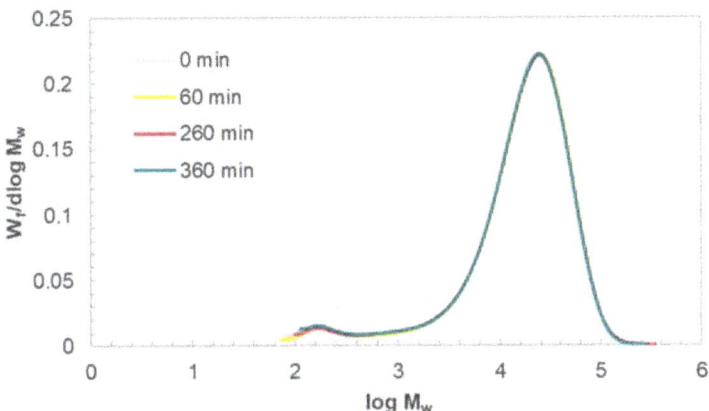

Figure 5. GPC chromatograms of the solid residue from sample B, laundered and cut, after 60, 260, and 360 min of hydrolysis, along with the reference sample, prior to hydrolysis.

Further, it confirms that the reaction proceeds predominantly through endwise peeling and is surface specific. Water and the hydrated OH^- catalyzing the hydrolysis have a low penetration into PET, also at the reaction temperature, 90 °C. At 90 °C, PET is still below the glass transition temperature, the temperature at which the polymer chains in the amorphous regions gain considerable mobility. However, even above the glass transition temperature, water penetration into the PET matrix is minimal, with typical water regains of below 0.5% [26]. The low penetration of the reaction medium into the polymer, together with the identical MWD over the course of reaction, confirms that the bulk of PET is not affected by hydrolysis but the reaction is surface specific.

Further, the microstructure of samples was assessed using WAXS (Figure 6). The peaks centered at 1.25, 1.63, and 1.82 refer to crystal planes (010), (−110), and (100) [33]. In terms of resolution, the peaks at higher q-number show a slight increase with hydrolysis time. This sharpening of peaks may be related to a slight increase in crystallinity. Similarly, to the GPC chromatograms, the curves are overlapping, indicating that the morphology is not significantly changed over the course of the reaction, in line with a surface specific reaction.

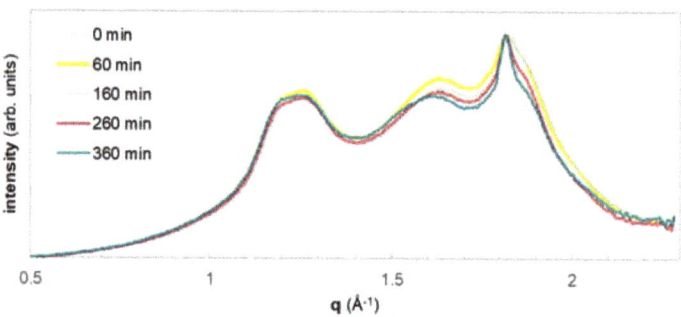

Figure 6. WAXS 1D spectra of the solid residue from sample B, laundered and cut, after 60, 160, 260, and 360 min of hydrolysis, along with the reference sample, prior to hydrolysis.

4. Conclusions

Depolymerization of textile PET by means of alkaline hydrolysis was studied in samples of different accessibility. Laundering of samples was found to generate a slight increase in the hydrolysis rate, in comparison to non-laundered samples, due to removal of surface finishing agents and processing aids. Mechanical disintegration, by means of shredding, had no impact on the rate of reaction. This is ascribed to the yarn structure where a low bulk density, originating from the use of a texturized yarn, ensures a high accessible area of the fabric even without disintegration. Mechanical pretreatment of PET textile to increase the accessibility prior to hydrolysis, hence, was shown to be unnecessary. The degradation mechanism of textile PET was found to correlate well with mechanisms earlier proposed for bulk PET. Neither GPC nor WAXS show any large changes in the MWD or morphology of the solid residues of samples over the course of reaction, indicating that the reaction is surface specific. The almost 100% yield of TPA at each sampling point during hydrolysis indicates that the reaction proceeds by peeling from the ends of the molecular chains, as has been previously shown for bulk PET. Similarly, NMR spectroscopy of the reaction filtrate show pure TPA and EG and no trace of oligomeric compounds.

Author Contributions: Conceptualization, J.B. and K.J.; Funding acquisition, H.d.l.M.; Investigation, A.P. and A.I.; Writing—original draft, A.P. and A.I.; Writing—review and editing, J.B., H.d.l.M. and K.J. All authors have read and agreed to the published version of the manuscript.

Funding: This research was funded by Södra Skogsägarnas Stiftelse för Forskning, Utveckling och Utbildning, grant number 2019-106.

Institutional Review Board Statement: Not applicable.

Informed Consent Statement: Not applicable.

Data Availability Statement: No data sets are made publicly available. However data can be obtained from the corresponding authors at any time.

Acknowledgments: Södra Innovation is acknowledged for fruitful discussions. Anja Lund, Simonetta Granelo, Jehona Sjöberg, Desiree Rex, and Angele Cruz are acknowledged for invaluable experimental help. We would also like to acknowledge the use of Chalmers Material Characterization Lab (CMAL), and the help from Michal Strach.

Conflicts of Interest: The authors declare no conflict of interest. The industrial reference group contributed to the formulation of the research question but had no role in the design, execution, interpretation, or writing of the study.

References

1. Exchange Textile. Preferred Fiber & Materials Market Report 2020. 2020. Available online: https://textileexchange.org/wp-content/uploads/2020/06/Textile-Exchange_Preferred-Fiber-Material-Market-Report_2020.pdf (accessed on 10 February 2022).
2. Manshoven, S.; Chistis, M.; Vercalsteren, A.; Arnold, M.; Nicolau, M.; Lafond, E.; Fogh, L.; Coscieme, L. *Textiles and the Environment in a Circular Economy*; European Topic Centre on Waste and Materials in a Green Economy: Mol, Belgium, 2019.
3. Sandin, G.; Peters, G.M. Environmental impact of textile reuse and recycling—A review. *J. Clean. Prod.* **2018**, *184*, 353–365. [CrossRef]
4. Güçlu, G. Alkyd resins based on waste PET for water-reducible coating applications. *Polym. Bull.* **2010**, *64*, 739–748. [CrossRef]
5. Kárpáti, L.; Fogarassy, F.; Kovácsik, D.; Vargha, V. One-pot depolymerization and polycondensation of PET based random oligo- and polyesters. *J. Polym. Environ.* **2019**, *27*, 2167–2181. [CrossRef]
6. Nikles, D.E.; Farahat, M.S. New motivation for the depolymerization products derived from poly(ethylene terephthalate) (PET) waste: A review. *Macromol. Mater. Eng.* **2005**, *290*, 13–30. [CrossRef]
7. Östlund, Å.; Wedin, H.; Bolin, L.; Berlin, J.; Jönsson, C.; Posner, S.; Smuk, L.; Eriksson, M.; Sandin, G. *Textilåtervinning—Tekniska Möjligheter och Utmaningar*; Naturvårdsverket: Stockholm, Sweden, 2015.
8. Palme, A.; Peterson, A.; de la Motte, H.; Theliander, H.; Brelid, H. Development of an efficient route for combined recycling of PET and cotton from mixed fabrics. *Text. Cloth. Sustain.* **2017**, *3*, 4. [CrossRef]
9. Barot, A.A.; Sinha, V.K. Chemical scavenging of post-consumed clothes. *Waste Manag.* **2015**, *46*, 86–93. [CrossRef]
10. Zhao, M.; Wang, X.; Yu, J. Preparation and characterization of poly(ethylene terephthalate) copolyesters and fibers modified with sodium-5-sulfo-bis-(hydroxyethyl)-isophthalate and poly(ethylene glycol). *J. Text. Inst.* **2016**, *107*, 1284–1295. [CrossRef]
11. Li, J.; Yang, Y.; Xiao, Y.; Tang, B.; Ji, Y.; Liu, S. Glucose-derived carbon nanospheres as flame retardant for polyethylene terephthalate. *Front. Mater.* **2021**, *8*, 539. [CrossRef]
12. Raheem, A.B.; Noor, Z.Z.; Hassan, A.; Abd Hamid, M.K.; Samsudin, S.A.; Sabeen, A.H. Current developments in chemical recycling of post-consumer polyethylene terephthalate wastes for new materials production: A review. *J. Clean. Prod.* **2019**, *225*, 1052–1064. [CrossRef]
13. O-Reilly, M.; Stubbe, J.A. PET polymer recycling. *Biochemistry* **2020**, *59*, 2316–2318. [CrossRef]
14. Manshoven, S.; Smeets, A.; Arnold, M. *Plastic in Textiles: Potentials for Circularity and Reduced Environmental and Climate Impacts*; European Topic Centre on Waste and Materials in a Green Economy: Mol, Belgium, 2021.
15. Smithers Market Report; *The Future of PET Packaging to 2025*; Smithers: Acron, OH, USA, 2020.
16. Kawai, F.; Kawabata, T.; Oda, M. Current state and perspectives related to the polyethylene terephthalate hydrolases available for biorecycling. *ACS Sustain. Chem. Eng.* **2020**, *8*, 8894–8908. [CrossRef]
17. Štrukil, V. Highly efficient solid-state hydrolysis of waste polyethylene terephthalate by mechanochemical milling and vapor-assisted aging. *ChemSusChem* **2021**, *14*, 330–338. [CrossRef] [PubMed]
18. Ügdüler, S.; Van Geem, K.M.; Denolf, R.; Roosen, M.; Mys, N.; Ragaert, K.; De Meester, S. Towards closed-loop recycling of multilayer and coloured PET plastic waste by alkaline hydrolysis. *Green Chem.* **2020**, *22*, 5376–5394. [CrossRef]
19. Payne, J.; Jones, M.D. The chemical recycling of polyesters for a circular plastics economy: Challenges and emerging opportunities. *ChemSusChem* **2021**, *14*, 4041–4070. [CrossRef]
20. Geyer, B.; Lorenz, G.; Kandelbauer, A. Recycling of poly(ethylene terephthalate)—A review focusing on chemical methods. *Express Polym. Lett.* **2016**, *10*, 559–586. [CrossRef]
21. Glaus, M.A.; Van Loon, L.R. Degradation of cellulose under alkaline conditions: New insights from a 12 years degradation study. *Environ. Sci. Technol.* **2008**, *42*, 2906–2911. [CrossRef]
22. López-Fonseca, R.; González-Marcos, M.P.; González-Velasco, J.R.; Gutiérrez-Ortiz, J.I. A kinetic study of the depolymerisation of poly(ethylene terephthalate) by phase transfer catalysed alkaline hydrolysis. *J. Chem. Technol. Biotechnol.* **2009**, *84*, 92–99. [CrossRef]
23. Aguado, A.; Martínez, L.; Becerra, L.; Arieta-araunabeña, M.; Arnaiz, S.; Asueta, A.; Robertson, I. Chemical depolymerisation of PET complex waste: Hydrolysis vs. glycolysis. *J. Mater. Cycles Waste Manag.* **2014**, *16*, 201–210. [CrossRef]
24. Carta, D.; Cao, G.; D'Angeli, C. Chemical recycling of poly(ethylene terephthalate) (PET) by hydrolysis and glycolysis. *Environ. Sci. Pollut. Res.* **2003**, *10*, 390–394. [CrossRef]
25. Kumar, S.; Guria, C. Alkaline hydrolysis of waste poly(ethylene terephthalate): A modified shrinking core model. *J. Macromol. Sci. Part A* **2005**, *42*, 237–251. [CrossRef]
26. Broadbent, A.D. *Basic Principles of Textile Coloration*; Society of Dyers and Colourists: West Yorkshire, UK, 2001.
27. Al-Sabagh, A.M.; Yehia, F.Z.; Eshaq, G.; Rabie, A.M.; ElMetwally, A.E. Greener routes for recycling of polyethylene terephthalate. *Egypt. J. Pet.* **2016**, *25*, 53–64. [CrossRef]
28. López-Fonseca, R.; González-Velasco, J.R.; Gutiérrez-Ortiz, J.I. A shrinking core model for the alkaline hydrolysis of PET assisted by tributylhexadecylphosphonium bromide. *Chem. Eng. J.* **2009**, *146*, 287–294. [CrossRef]
29. Mishra, S.; Goje, A.S. Chemical recycling, kinetics, and thermodynamics of alkaline depolymerization of waste poly(ethylene terephthalate) (PET). *Polym. React. Eng.* **2003**, *11*, 963–987. [CrossRef]
30. Yoshioka, T.; Okayama, N.; Okuwaki, A. Kinetics of hydrolysis of PET powder in nitric acid by a modified shrinking-core model. *Ind. Eng. Chem. Res.* **1998**, *37*, 336–340. [CrossRef]

31. Yoshioka, T.; Motoki, T.; Okuwaki, A. Kinetics of hydrolysis of poly(ethylene terephthalate) powder in sulfuric acid by a modified shrinking-core model. *Ind. Eng. Chem. Res.* **2001**, *40*, 75–79. [CrossRef]
32. Lykaki, M.; Zhang, Y.Q.; Markiewicz, M.; Brandt, S.; Kolbe, S.; Schrick, J.; Rabe, M.; Stolte, S. The influence of textile finishing agents on the biodegradability of shed fibres. *Green Chem.* **2021**, *23*, 5212–5221. [CrossRef]
33. Chen, S.; Xie, S.; Guang, S.; Bao, J.; Zhang, X.; Chen, W. Crystallization and thermal behaviors of poly(ethylene terephthalate)/bisphenols complexes through melt post-polycondensation. *Polymers* **2020**, *12*, 3053. [CrossRef]

Article

Chemical Recycling of a Textile Blend from Polyester and Viscose, Part I: Process Description, Characterization, and Utilization of the Recycled Cellulose

Anna Peterson [1], Johan Wallinder [2], Jenny Bengtsson [1], Alexander Idström [3], Marta Bialik [2], Kerstin Jedvert [1] and Hanna de la Motte [1,*]

1 RISE Research Institutes of Sweden, Argongatan 30, Box 104, SE-431 22 Mölndal, Sweden; anna.peterson@ri.se (A.P.); jenny.bengtsson@ri.se (J.B.); kerstin.jedvert@ri.se (K.J.)
2 RISE Research Institutes of Sweden, Drottning Kristinas väg 61, Box 5604, SE-114 86 Stockholm, Sweden; johan.wallinder@ri.se (J.W.); marta.bialik@ri.se (M.B.)
3 Department of Chemistry and Chemical Engineering, Chalmers University of Technology, SE-412 96 Gothenburg, Sweden; idstrom@chalmers.se
* Correspondence: hanna.delamotte@ri.se

Abstract: Material recycling requires solutions that are technically, as well as economically and ecologically, viable. In this work, the technical feasibility to separate textile blends of viscose and polyester using alkaline hydrolysis is demonstrated. Polyester is depolymerized into the monomer terephthalic acid at high yields, while viscose is recovered in a polymeric form. After the alkaline treatment, the intrinsic viscosity of cellulose is decreased by up to 35%, which means it may not be suitable for conventional fiber-to-fiber recycling; however, it might be attractive in other technologies, such as emerging fiber processes, or as raw material for sugar platforms. Further, we present an upscaled industrial process layout, which is used to pinpoint the areas of the proposed process that require further optimization. The NaOH economy is identified as the key to an economically viable process, and several recommendations are given to decrease the consumption of NaOH. To further enhance the ecological end economic feasibility of the process, an increased hydrolysis rate and integration with a pulp mill are suggested.

Keywords: textile recycling; textile blend; viscose; polyester; industrial process layout

1. Introduction

World fiber production has seen a continuous increase over the last decades, and the current world fiber market reaches well above 100 million tons of produced fibers per year [1], where the majority is used for apparel and home textiles. Consequently, the amount of disposed textiles is increasing, and there is an urgent need for improved handling of end-of-life textiles and textile recycling [2]. One major challenge considering the recycling of textiles is the abundance of multi-material textiles, i.e., fabrics composed of two or more different types of polymers. Mechanical recycling of such multi-material textiles produces materials of inferior properties, due to the distinct properties and processing needs of the different fiber types. There is currently a lack of recycling options for these textiles, which results in them being predisposed for landfill or incineration.

The mixture of polyester (PET) and cellulose fibers, especially cotton, is common, due to the beneficial effect of combining PET, which is durable and inexpensive, with cellulose fibers, which have good water absorbency, thereby contributing to the comfort of the fabric. However, to recycle this type of material, a separation is required for further valorization of the two components. Various routes of cellulose/PET separation have been investigated, i.e., through degradation, depolymerization, or dissolution of at least one of the two components.

PET can be maintained in its polymeric form through selective dissolution [3–5] or hydrolysis [6,7] of cotton. However, the molecular weight of PET is decreased in a recycled fabric compared to a virgin fabric due to laundering and wear. The lowered molecular weight of PET may result in recycled PET fibers with inferior mechanical properties, or even make the melt–spinning process impossible. In contrast, if PET is depolymerized, PET monomers are obtained, which, after suitable purification, may be used in the production of new PET [8]. Cotton from cellulose/PET textiles is currently recycled into dissolving pulp on a commercial or demo scale by actors such as OnceMore, Ambercycle, and WornAgain.

Simultaneous depolymerization of PET and recovery of the cotton fibers through alkaline hydrolysis has previously been successfully demonstrated [9,10]. While the aqueous alkaline environment causes hydrolytic cleavage of the ester bond of PET, cotton is relatively resistant to alkali degradation [11]. In particular, in the presence of a phase transfer catalyst, the depolymerization of PET in 10% NaOH at 90 °C is fulfilled within less than 1 h, with limited degradation of the cotton fraction [9]. However, whether alkaline hydrolysis can be successfully applied to fabrics containing PET blended with other types of cellulose fibers, e.g., regenerated fibers such as viscose or Lyocell, remains to be investigated.

Cotton and regenerated cellulosic fibers differ in several aspects, which will dictate how the cellulose is affected under alkaline conditions. The first distinction is in the ultrastructure of fibers, with the presence of a primary cell wall in cotton fibers, a structural feature that regenerated fibers lack [12]. Compared to regenerated cellulose fibers, the cellulose in cotton is of much higher molecular weight. Cotton has a typical average degree of polymerization (DP) of 2000 [13], Lyocell in the range of 400–700 [14], and viscose even lower, between 250–400 [13]. Further, the crystallinity varies between fibers—for native cotton, it is around 70%, while for viscose about 25–30%. The crystallites in regenerated fibers are also much smaller compared to cotton, and have a lower degree of orientation [13].

At the relevant process conditions, a NaOH concentration of a maximum of 10 wt% and temperatures below 100 °C, cellulose is known to undergo peeling reactions. Endwise peeling results in the removal of anhydroglucose units from the reducing end of the cellulose chain. Peeling will continue until the competitive stopping reaction occurs, and, depending on the conditions, i.e., alkalinity and temperature, peeling may proceed for 50–60 units [15]. Regenerated cellulosic fibers, having a lower DP than cotton, are potentially more prone to these peeling reactions.

A recent life-cycle analysis study concluded that there are potential environmental benefits in using cotton rendered from separation of cotton/PET through alkaline hydrolysis as a raw material in viscose or Lyocell production [16]. For regenerated fibers, the DP might be too low after separation, and, consequently, the cellulose fibers would not be suitable as raw material for conventional fiber-to-fiber recycling, but may be recovered as other products [17]. One option could be to produce cellulose nanocrystals from the cellulose fraction, which has been demonstrated using both cotton and cotton/PET blends as a raw material [18,19]. Furthermore, the complete depolymerization of cellulose into glucose allows for the production of specialty chemicals or biofuels [20]. Such processes to valorize degraded cellulosic textiles have already been proposed, commonly through a pretreatment of the cellulose, followed by enzymatic hydrolysis. Alkaline pretreatment with bases such as sodium hydroxide [21] and sodium carbonate [22] has been described, and could favorably be combined with alkaline hydrolysis of cellulose/PET textile blends.

The properties of the cellulose fraction, after hydrolysis and separation of a viscose/PET textile blend, dictate which recycling pathways will be most promising. Thus, in this study, alkaline hydrolysis of viscose/PET blends was performed, and the cellulose fibers were evaluated with respect to yield, molecular weight, and fiber morphology. Furthermore, a process layout for an industrial upscaling of the process for cellulose/PET separation via alkaline hydrolysis was constructed based on the experimental results. The process layout was constructed in the form of a mass balance model using the process simulation tool WinGEMS, and can serve as a basis to indicate the economic and ecological potential of an upscaling of the proposed hydrolysis. We believe that applied studies on material recycling can benefit largely from a close collaboration between experimental work and techno–economical and life-cycle analysis. It allows the direction of the experimental work to better favor an economically and ecologically viable process, a factor that is often overlooked. Any recycling option must be better than the use of virgin materials, i.e., have a lower climate impact, and be economically feasible. In this work, an upscaled industrial process layout is presented, which identifies the areas of the proposed process that require further optimization. Moreover, it does not put hard numbers on the operations, a practice that might prematurely disregard a promising recycling option due to a lack of reliable data in the early stages of process development.

2. Experimental Section

2.1. Materials

For this study, three types of never-used viscose were used: a commercially available garment with a blended fabric containing 70% viscose and 30% polyester, a mixture of 70% viscose and 30% polyester filaments, and a pure viscose fabric supplied by a Swedish fashion company. The two different fabrics were either used as received or washed, and subsequently cut into approximately 1×1 cm pieces. Laundering was performed at 60 °C, according to Swedish standard SS-EN ISO 6330:2012. A portion of the laundered fabric was also shredded (New Shunxing, NSX-QT 310). Moreover, a neat polyester fabric was laundered and cut in the same manner as the neat viscose fabric. Sodium hydroxide (NaOH, 50%) and sulfuric acid (H_2SO_4, reagent grade, 95–97%) were obtained from Sigma-Aldrich, and acetic acid (glacial, 99%,) was obtained from Fisher Scientific, and used as received.

2.2. Methods

2.2.1. Hydrolysis

Alkaline hydrolysis of viscose/PET, neat viscose, and neat PET samples was performed at a solids-to-liquids ratio of 1:100 in aqueous NaOH. The NaOH concentration was 5 wt%. The aqueous NaOH was heated to 90 °C before adding the oven-dried (2 h, 105 °C) sample to the reaction vessel. Hydrolysis was performed for the selected time (60–1440 min). The reaction was quenched by immersing the reactor in an ice bath. Post-reaction, the solid residue was separated from the reaction solution via filtration, using a tight knit PET–wire fabric as the barrier. The solids were neutralized in 5% acetic acid, and thereafter washed with water and dried at 105 °C for 4 h to calculate the gravimetric yield. The filtrate was kept in the freezer until further analysis.

2.2.2. Composition

The composition of the mixed viscose/PET fabric was probed via selective dissolution of viscose in cupriethylenediamine (CED), whereafter the weight of the solid PET residue was determined.

2.2.3. Intrinsic Viscosity

Limiting viscosity of the cellulose samples was measured after dissolution in CED, according to ISO 5351. The degree of polymerization was calculated from the correlation formulated by Immergut et al. [23], as cited in SCAN-C 15:62 ($DP^{0.905} = 0.75[\eta]$, η given in cm^3/g). The correlation is known to be flawed, but is commonly used, which facilitates comparisons to other studies.

2.2.4. WAXS

The crystallinity of viscose was measured by wide angle X-ray scattering (WAXS) at Chalmers University of Technology (Mat:Nordic, SAXSLAB) with a 0.9 mm beam diameter, and a Rigaku 003+ high brilliance microfocus Cu-radiation source at 130 mm distance, with the sample irradiated over 20 min. The intensity was normalized with transmission.

2.2.5. FTIR Spectroscopy

Attenuated total reflectance Fourier transform infrared (ATR–FTIR) spectroscopy was performed on a Bruker Tensor 27 equipped with the Specac Golden Gate ATR accessory.

2.2.6. NMR Spectroscopy

The purity of terephtalic acid (TPA) obtained from PET depolymerization was determined via nuclear magnetic resonance (NMR) spectroscopy. A small portion of the filtrate from the sample of Fabric70/30 after hydrolysis for 360 min was acidified with H_2SO_4 (4M) to precipitate TPA. TPA was dissolved in 5% $NaOH/D_2O$ solution prior to analysis. The NMR measurements were conducted on a Varian 400-MR spectrometer operating at 9.4 T, equipped with a OneNMRProbe. The ^1H-NMR spectrum parameters included a 5 µs ^1H-detection pulse, 2.5 s acquisition time, 2 s recycle delay, and 32 scans. The samples were studied using D_2O as a solvent, and the chemical shifts were referenced to the residual solvent signal.

2.2.7. Upscaled Process Layout

The proposed upscaled process layout was based on a mass and energy balance steady-state model constructed using the process simulation tool WinGEMS (Valmet®, Espoo, Finland). This modular simulation tool is specifically developed for the pulp and paper industry, and handles the modeling of suspensions with ease. The first step in creating the mass and energy balance model involves constructing a hypothetical process flowsheet, using the so-called blocks in the WinGEMS graphical environment. These are modular (drag and drop) and pre-coded for a vast set of process operations, spanning simple splitting and mixing to heat exchangers, evaporators, and washers. Moreover, the stream structure and its components need to be defined by the user, as do the necessary chemical reactions. The basic model for a chemical reaction is a simple stoichiometric conversion whose extent is also defined by the user, while the actual thermodynamic aspects are usually handled outside of the simulation environment using other, more rigorous tools. For the establishment of energy balances, WinGEMS has built-in expressions for estimating the specific enthalpies of the streams. For liquid streams, the enthalpy is a function of temperature, dissolved and suspended solids, and the specific heat of water and solids. The enthalpies of steam streams is determined by a built-in steam table (function of temperature and pressure). The amount of heat from reactions had to be manually calculated and used as an input for the heat exchanger blocks to estimate changes in the stream temperatures. In this study, the procedures employed within the experimental work, including the chemical additions, solids-to-liquids ratio, etc., as well as the experimental results (reaction yields, product purities, etc.), were used as a basic input for the model around the hydrolysis and precipitation stages. The remaining stages in the process layout were proposed based on the general engineering know-how regarding standard industrial processes (washing, heat exchangers, etc.), experiences from similar studies, reasonable simplifications and assumptions (discussed in detail in Section 3), as well as the basic principles of green chemistry. The results are hence to be seen as a first iteration of a possible upscaling of the process. The overall objective for the upscaling development was to obtain a hypothetical process layout which was as efficient as possible from the perspective of chemicals and energy use, and subsequently suggest improvements that would help to make it economically and environmentally sustainable.

3. Results and Discussion

3.1. Experimental Study

Alkaline hydrolysis of blends of viscose/PET fibers was studied for two different cases: a 70/30 mix of pure viscose and pure PET filaments (Filament70/30), and a knitted viscose/PET fabric with a 70/30 composition (Fabric70/30) (Table 1). The composition of the mixed fabric was confirmed via selective dissolution of viscose in CED and determination of the weight of the solid PET residue. The choice of samples allows to study the efficiency of the separation for samples with different accessibilities, i.e., a mix of filaments relative to a knit fabric from mixed yarns. Further, alkaline degradation of viscose in terms of mass loss, DP, and crystallinity was studied for a neat viscose fabric. The accessibility of neat viscose to alkaline degradation was studied for a 100% viscose fabric (Viscose100) prepared in three different ways: never laundered, cut to 1 × 1 cm pieces (Viscose100a); laundered, cut to 1 × 1 cm pieces (Viscose100b); and laundered, shredded to display the individual fibers (Viscose100c). Finally, the hydrolysis of a knitted PET fabric, which was laundered and cut to 1 × 1 cm pieces, was studied to produce the data used in the upscaled process model.

Table 1. Description of viscose-containing samples.

Sample	Composition PET/Viscose	Form	Intrinsic Viscosity (T_0) (mL g^{-1})
A	0/100	Fabric	
A1	0/100	Never laundered, cut	167
A2	0/100	Laundered, cut	170
A3	0/100	Laundred, shredded	171
B	30/70	Filaments	196
C	30/70	Garment from blended fibers	n.d.

Mild reaction conditions (90 °C, 5 wt% NaOH) and reaction times between 60–1440 min were chosen, as the DP of cellulose has been shown to decrease with increasing temperature and NaOH concentration during hydrolysis [9]. For closed loop-recycling, i.e., fiber-to-fiber recycling of regenerated fibers, it is of the utmost importance to preserve the molecular weight of the cellulose fraction in the mixed fabric. Simultaneously, to allow for separation of the blended textile, the reaction conditions should allow for full PET depolymerization. Furthermore, a reaction below the boiling point of the aqueous solution eliminates the need for a pressurized system and consumes less energy. The reaction yields a solid cellulose residue, and the PET monomers terephtalic acid (TPA) and ethylene glycol (EG) (Figure 1).

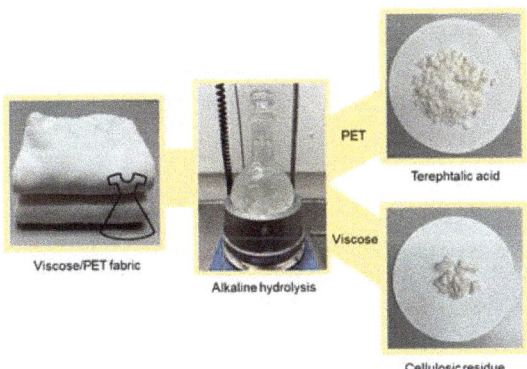

Figure 1. Schematic of the reaction.

The mass losses of samples Filament70/30 and Fabric70/30 were comparable after 60 min of hydrolysis (Figure 2a), and were also in the same range as Viscose100 (Figure 2b). This implies that no major hydrolysis of PET occurs within the first 60 min of reaction, but the mass loss originates predominantly from viscose. We ascribe this to fast degradation of low molecular weight fragments and accessible regions of the cellulose in viscose. At longer reaction times, the mixed yarns and knitted structure of Fabric70/30 does decrease the hydrolysis rate and, consequently, the mass loss compared to Filament70/30 is also decreased. Even after 24 h of hydrolysis, PET removal is not complete in Fabric70/30, indicated by the <30% mass loss. This is confirmed by FTIR analysis, where the ester peak at 1705 cm^{-1}, significant for PET, is still visible after 24 h of hydrolysis (Figure 3).

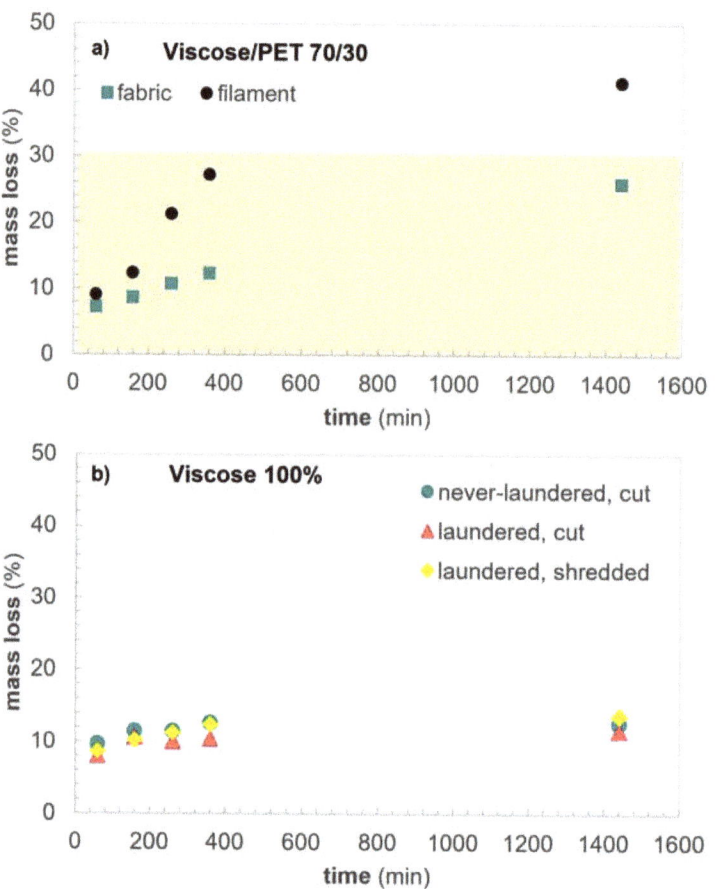

Figure 2. Mass loss during hydrolysis of (**a**) blends of PET and viscose and (**b**) 100% viscose fabric. The yellow part in Figure 2a represents the wt% of PET in the samples.

In contrast, Filament70/30 shows a 40% mass loss, indicating full depolymerization of PET and an additional 10% mass loss from cellulose, consistent with the mass loss in a neat viscose sample after 24 h reaction. Textile construction of a blended fabric, hence, has a substantial impact on the hydrolysis rate of PET. Contrary, for Viscose100 there is no significant difference between the mass losses of samples of different accessibility. Laundering the sample at 60 °C removes surface finishing and other processing aids remnant on the fabric which increase the exposure to the alkaline environment during hydrolysis, while shredding increases the surface area of fabrics. Interestingly, most of the mass loss is seen during the first 60 min of reaction for all viscose samples (Figure 2b), where the never-laundered sample displays a 10% weight loss and the laundered samples, cut or shredded, 8 and 9%, respectively. After 24 h the weight loss has increased to 13% for the never-laundered sample and 12 and 14% for the laundered cut and shredded samples, respectively. Thus, more than 50% of the total mass loss occurs during the early stages of reaction.

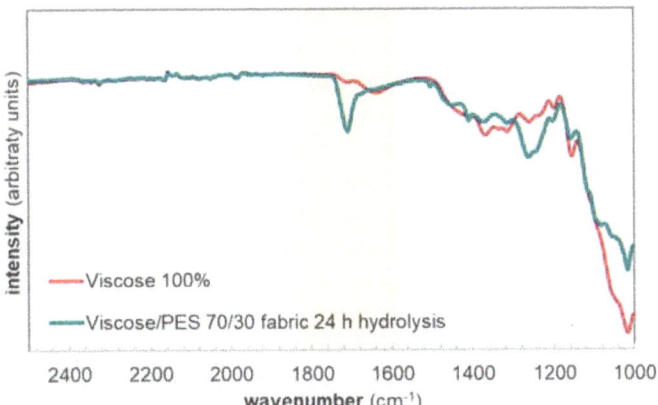

Figure 3. FTIR spectra of the viscose/PET fabric after 24 h of hydrolysis, and a reference 100% viscose fabric. Highlighted is the peak at 1705 cm^{-1}, significant for the ester bonds in PET.

Characterization of Reaction Products

The mass loss during the first 60 min of reaction is accompanied with a significant increase in the crystallinity of the Viscose100b sample (Figure 4). This is signified by a sharpening of the peaks centered around 2θ values of 12 and 20. These peaks have previously been assigned to cellulose II, the crystalline structure of regenerated cellulosic fibers [24]. The supramolecular structure of cellulose is decisive to the kinetics of all degradation reactions [25], and a higher rate of alkaline peeling is expected in disordered, compared to crystalline, regions. The increase in crystallinity is in line with our postulate that the mass loss during the first hour of hydrolysis is due to the removal of low molecular weight fragments and accessible regions of the viscose, by means of peeling reactions.

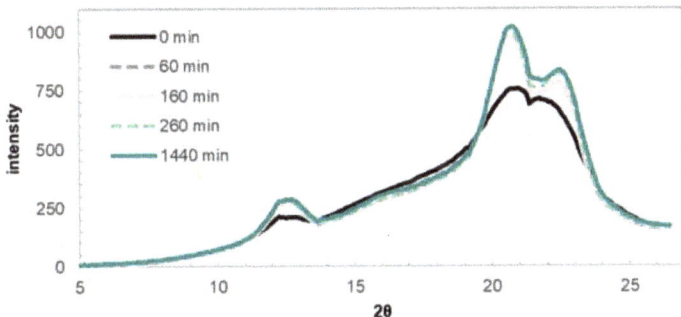

Figure 4. WAXS 1D spectra of sample Viscose100b after 60, 160, 260, and 1440 min of reaction time, as well as the reference sample prior to immersion in hot alkali.

The purity of precipitated TPA from the viscose/PET fabric was determined via NMR spectroscopy, with a commercial TPA as the reference. The spectrum of the precipitate shows a distinct singlet at 7.58 ppm, which is assigned to pure TPA (Figure 5). A zoom-in of the baseline around the signal is shown as an insert. No other signals than the ^{13}C-satellites from singlet A (in the insert marked with an *) can be seen in the zoom-in, proving that no residues of other reaction products stemming from incomplete depolymerization of PET are present in the precipitate. This is in accordance with the literature, where hydrolysis of PET is described as taking place at the external surface of the solid-state PET, with the major decomposition reaction occurring at the ends of the polymer chains [26].

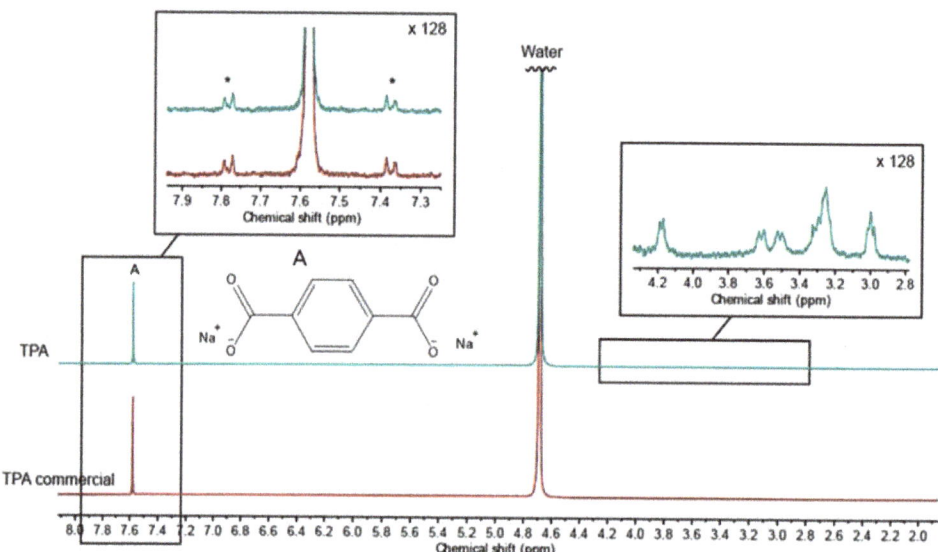

Figure 5. ^{1}H-NMR spectra of TPA precipitated from the reaction filtrate, along with a commercial TPA reference. The singlet from TPA is marked with an A, and in the zoom-in of this signal, the ^{13}C-satellites from A is marked with an *. Zoom-in around 3.4 ppm show residual viscose degradation products.

Around 3.4 ppm, the experimentally obtained TPA displays some additional small peaks, characteristic for protons in the vicinity of oxygen, as seen in the insert. These are assigned to traces of viscose degradation products, such as D-glucoisosaccharinic acids, the predominant product of peeling reactions during alkaline conditions [15].

Having confirmed the purity of TPA obtained from hydrolysis of PET from viscose/PET, further characterization of the cellulosic residue was undertaken. The molecular weight of cellulose is commonly approximated by the intrinsic viscosity (IV) of a cellulose solution in cupri–ethylene diamine (CED). Prior to alkaline treatment, the viscose filaments in sample Filament70/30 had an IV of 200 mL g^{-1}, while the IV of Viscose100 was slightly lower, 170 mL g^{-1} (Table 1). The IV of Fabric70/30 could not be measured, due to degradation. Cellulose is known to degrade in CED, i.e., the solvent used for measuring IV. Even when finely dispersed in the solvent, complete dissolution of the cellulose part of the viscose/PET fabric could not be achieved before degradation occurred, hence the obtained IV values were not reliable, and are therefore not reported. However, data on PET depolymerization from Fabric 70/30 (Figure 3) indicate that a reaction time of at least 24 h is needed for full PET depolymerization under the present, mild reaction conditions. We assume that the degradation of viscose will not be significantly impacted by the presence of polyester in the blended fabric, and may be approximated by the IV of neat viscose samples. For all neat viscose samples, the IV decreased with reaction time, and measured between 130–150 mL g^{-1} after 24 h (Figure 6). The IV may be approximatively converted to DP, using the relationship formulated by Immergut [23] (c.f. Experimental section):

$$DP^{0.905} = 0.75[\eta] \tag{1}$$

where [η] is the intrinsic viscosity in mL g^{-1}. After 24-h reaction, the IV is decreased by between 20–50 mL g^{-1}, which, using Equation (1), corresponds to 20–55 glucose units. The decrease in DP is in line with a peeling reaction, which typically proceeds for up to 50–60 units before a stopping reaction occurs [15]. The final IV of samples is too low for applications in commercial regenerated cellulose fibers or cellulose derivates, such as cellulose ethers, nitrates, and acetates, which typically lies in the range of 400–600 mL g^{-1} [27]. However, low molecular weight polysaccharides can find other uses, e.g., by hydrolysis into nanocrystalline cellulose [28] or in sugar platforms for further valorization into specialty chemicals and biofuels [29].

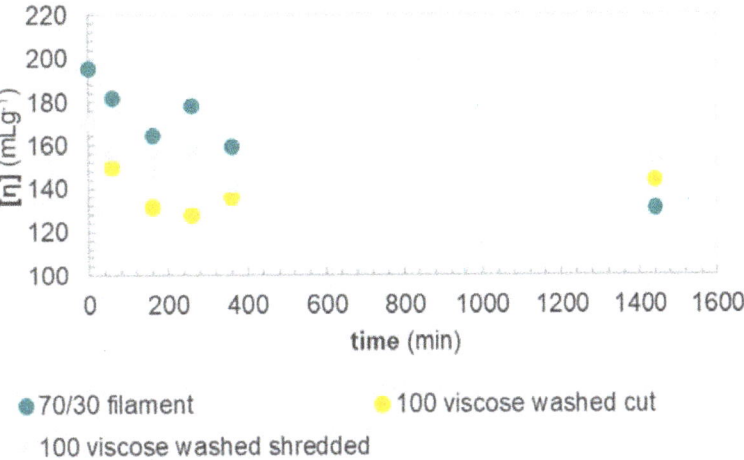

Figure 6. Intrinsic viscosities of samples as a function of hydrolysis time.

3.2. Upscaled Process Layout

Based on the relevant experimental results in the section above, as well as a number of assumptions, a process layout for a greenfield standalone industrial upscaling of the viscose/PET separation process was constructed. The process layout aims at providing guidelines on the aspects of the proposed industrial process, which require more development and optimization for the whole process to become viable from an economic and environmental point of view. In essence, the process layout should be seen as a first step on a necessary path of various iterations between process development and experimental work to lift the process to higher technology readiness levels (TRLs). With respect to the early experimental results of the project, rather than conducting a conventional quantitative techno–economic evaluation, which may result in irrelevant results at this stage, this study emphasizes the qualitative measures to be implemented in future projects. The ultimate objective of the process is to completely hydrolyze PET and separate its monomers, TPA and EG, while keeping the viscose fraction as intact as possible. The PET monomers could be sold for production of PET, or used as a platform chemical. The IV of viscose afterhydrolysis (Figure 6) is likely too low for fiber-to-fiber recycling, and while other outputs are proposed, these are currently not available on an industrial scale, and their economic value is difficult to estimate.

3.2.1. Use of Experimental Data

The yields used in the hydrolysis stage in the model were based on experiments with pure fabric fractions of PET and viscose, respectively, as demonstrated in Table 2. The composition of the inlet textile in the model is 70/30 viscose/PET. The textile-to-solvent ratio and NaOH concentration were kept at 1:100 and 5 wt%, respectively, the same as in the experimental study.

Table 2. Experimental values of mass loss from hydrolysis of neat viscose or neat PET fabric used in the model.

Time of Hydrolysis (min)	Viscose Mass Loss (%)	PET Mass Loss (%)
60	8.0	15.6
160	10.6	35.1
260	10.0	56.2
360	10.4	68.5
1440	11.6	99.8

Since it was difficult to obtain a reliable alkali consumption from the experimental results, a theoretical consumption based on the respective mass losses of PET and viscose was assumed. The hydrolysis of PET into its monomers was modeled according to Equation (2). It was assumed that the PET fabric is composed of only poly(ethylene terephthalate) (PET), i.e., the hydrolysis of PET requires two moles of NaOH per repeating unit. While this is true for most textile PET, textile terminology does not distinguish between PET and other types of polyesters. Furthermore, according to the same textile terminology, up to 15% of non-polyester copolymers are allowed in the polymer chain [30]. For viscose, it was approximated that all of the degraded viscose would be recovered as glucose, and the formation of each glucose molecule would consume one OH^- ion.

$$PET_n + 2NaOH \rightarrow TPA_{salt} + EG \tag{2}$$

The consumption of sulfuric acid in precipitation of TPA was based on the stoichiometric reactions in Equations (3)–(5). TPA has a pKa of 3.5, and sulfuric acid is consumed in the acidification of the filtrate as well as the protonation of TPA. The consumption of sulfuric acid required to reach a pH of 3.5 was estimated by a simulated titration in OLI. It was assumed that the complete precipitation of TPA took place in the model.

$$H_2SO_4 + 2\, NaOH \rightarrow Na_2SO_4 + 2\, H_2O \tag{3}$$

$$H_2SO_4 + 2\, H_2O \rightarrow 2\, SO_4^{2-} + 2\, H_3O^+ \tag{4}$$

$$H_2SO_4 + TPA_{salt} \rightarrow Na_2SO_4 + TPA_{acid} \tag{5}$$

3.2.2. Proposed Process Layout

With the exception of the experimental data presented in the previous section, all other process stages and values in the process layout originate from assumptions and estimates. Figure 7 displays the first proposed process layout for an upscaling of the viscose/PET separation process.

The textile feed is first mixed with the alkaline solvent before it is heated via two indirect heat exchangers to the temperature of the hydrolysis reactor. The heat of reaction from the neutralization of alkali (~57 kJ/mol NaOH) is used to heat exchange the textile–solvent feed in the first heat exchanger. The heat of reaction was calculated from the standard enthalpies of formation of H_2SO_4 (aq), NaOH, Na_2SO_4, and H_2O (l) [31]. External heating is required to ramp up the textile–solvent feed to the temperature of the hydrolysis reactor. A natural gas burner, operating at 3 bar, supplies the second heat exchanger with saturated steam. A gas burner was chosen above other technologies (e.g., biomass boiler, electric boiler) due to its simplicity, as well as the lower investment and operating costs. However, this means that the process must be placed in the proximity of a chemical cluster, with access to a natural gas infrastructure. The modeled residence time in the hydrolysis reactor, and, consequently, the yield, can be any of the ones presented in Table 2. Following the hydrolysis reactor is a simple dewatering press (without a wash liquor), from which the outlet consistency is 30%. It is assumed that the carryover, i.e., the liquid following the viscose fraction, has the same composition as the filtrate from the press. Next, two counter-current wash presses with the same discharge consistency purify the viscose fraction from residual alkali, dissolved PET monomers, etc. The assumed displacement ratio (washing efficiency) and dilution factor (mass of excess filtrate per mass of viscose) is ~0.5 and ~5 ton/ton viscose, respectively. Both the displacement ratio and the dilution factor are typical parameters applied in pulp mills during the washing of the pulp. The filtrate from the wash presses is subsequently mixed with the filtrate from the dewatering press. To make the viscose fraction easier to transport, a pressing and drying section increases the viscose consistency to 90%. The gas burner also supplies the drying section with the necessary heat.

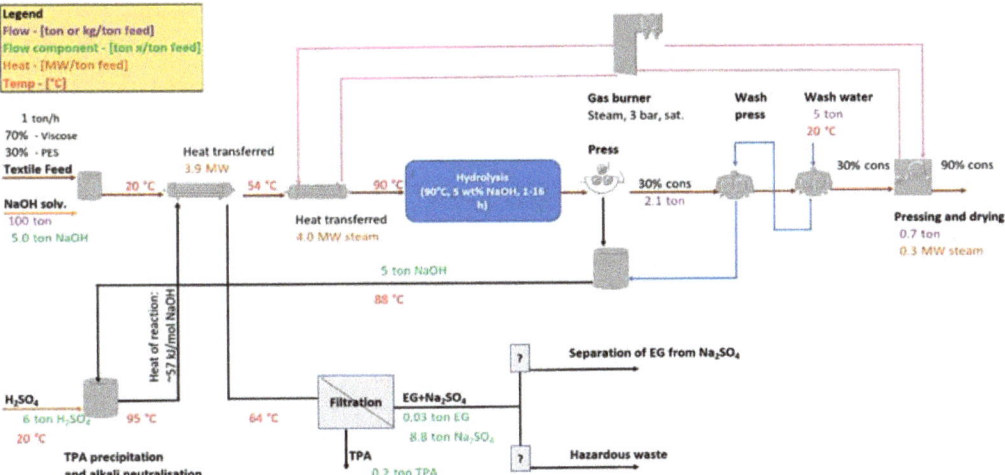

Figure 7. The proposed process layout for the viscose/PET separation process via alkaline hydrolysis. Purple values indicate total flow, green indicates component flow in the stream, brown indicates heat transfer, and red indicates the temperature. The values are to be seen as indicative and, in this case, pertain to a hydrolysis residence time of 24 h.

The filtrate from the dewatering press is sent to a vessel for precipitation of TPA with sulfuric acid. As mentioned previously, the reaction of neutralizing alkali with sulfuric acid is heavily exothermic, and the excess heat is used to pre-heat the feed. Subsequent washing and filtering of the outlet stream is then performed to separate the TPA from the process. The remaining aqueous phase contains EG, degraded viscose, and Na_2SO_4, formed during the neutralization of the residual alkali. As indicated by Figure 7, the most suitable fate of this stream is still uncertain, and will be discussed further at a later stage.

Below is a list of additional assumptions in the process layout:

- There is no heat loss from the heat exchangers;
- The hydrolysis reactor is assumed to be well-insulated, so that the temperature of 90 °C can be held constantly during the reaction, with no need for supporting heat;
- The viscose fraction is still pumpable at a consistency of 30%;
- For the heating of water, the specific heat capacity is 4.2 kJ kg K^{-1}.

3.2.3. Suggestions for Process Optimization

The proposed process layout in Figure 7 is, in essence, a direct translation of the laboratory procedures, and results in an upscaled process; consequently, certain areas and aspects of the proposed process need further elaboration to make them more efficient and applicable. Minimizing the basic operational expenditures (chemicals, energy etc.) is one of the most apparent and necessary actions to focus on. Investigation into the most suitable management of the residual EG/Na_2SO_4 fraction, as well as the best application for the viscose fraction, would also be required to further develop the process.

The basic operational expenditures of the process are NaOH, natural gas, and H_2SO_4. In this process, the use of NaOH has been identified as the weakest point. Since NaOH is a relatively expensive chemical (~550 EUR/ton [32]), due to the high power consumption during its production, and, given its prominent role in the proposed process, efficient use of NaOH is key to the process economy. 5 tons NaOH/ton of textile feed is fed to the current process to reach the pH at which the reaction is assumed to proceed most efficiently. However, even in the process case with the highest alkali consumption, i.e., 24 h of hydrolysis (Table 2), the amount of residual alkali leaving the reactor is still as much as ~4.85 ton/ton of textile feed, i.e., only a fraction of this valuable chemical is irreversibly consumed. In the proposed process, this means that a massive amount of unused alkali, a value of around 2670 EUR/ton textile, follows the filtrate to the TPA precipitation tank, and is neutralized into Na_2SO_4. Not only is this an inefficient use of NaOH, but it is also an inefficient use of H_2SO_4 which is used for neutralization. At least three robust process improvements, to be evaluated separately and subsequently in combination, to increase the NaOH economy have been identified:

1. Lower the textile-to-solvent ratio during hydrolysis significantly from 1:100. Assuming the same alkali consumption for a textile-to-solvent ratio of 1:10 would mean an input of 0.50 ton NaOH/ton textile, a residual alkali of 0.35 ton/ton textile, and a value of ~193 EUR/ton textile that is neutralized. Initial experimental results show that wetting of the cut fabric is possible down to a textile-to-solvent ratio of 1:15, and that full depolymerization of neat PET textile is achieved after 24 h of hydrolysis at this ratio.
2. Lower the concentration of NaOH in the solvent from 5 wt%. Similar to the reduction of the textile-to-solvent ratio, this process change must also be experimentally evaluated to determine the lowest possible NaOH concentration without the risk of decreased yields.
3. Separate the TPA salt in the filtrate from the dewatering press before precipitation (Figure 7) with a nano filtration (NF) membrane and/or ion exchange technology. By doing so, the permeate, with most of the residual alkali, could be recycled, and mixed back in with the textile feed. There are pH-stable NF membranes that appear as promising candidates which are able to withstand the alkaline filtrate, with a cutoff down to 200 Dalton [33]. The TPA salt (TPA-Na_2) has a molecular weight of 210 g/mol, and might possibly be effectively separated in the NF membrane, although this should also be verified experimentally.

For the third process improvement listed above, a new iteration of the process layout was created (Figure 8). In it, it is assumed that an effective recirculation of 90% of the residual alkali with the aid of a NF membrane is possible. It is assumed that 10% of the total flow follows the retentate with up-concentrated TPA. Compared to Figure 7, the required makeup of fresh NaOH decreases by 87%, and the charge of H_2SO_4 decreases by 88%. The energy consumption also decreases by 72%, since the recycling of the high temperature residual alkali stream, at 87 °C, drastically lowers the need for external heating for the ramp up to 90 °C.

As mentioned previously, it will be necessary to identify treatment options for the residual EG/Na_2SO_4 fraction after TPA precipitation. Within the frames of this work, it has not yet been clarified whether it is possible to separate EG from Na_2SO_4. Although commercially available technologies for glycol recovery do exist, preliminary discussions with one of the providers of such a technology (cleantech company Recyctec [34]) indicate that the EG/Na_2SO_4 stream in question might be difficult to process due to its high conductivity. Furthermore, the amount of EG in the stream is relatively small, and it might not be economically viable to separate it, but perhaps it could be for Na_2SO_4, which dominates the composition of the stream. If it is suspected that the fraction cannot be separated, the question remains whether to regard it as hazardous waste, or if it can simply be treated in a wastewater treatment plant. A recycling company [35], which was contacted, concluded that a detailed composition of the stream would need to be available for the proper classification.

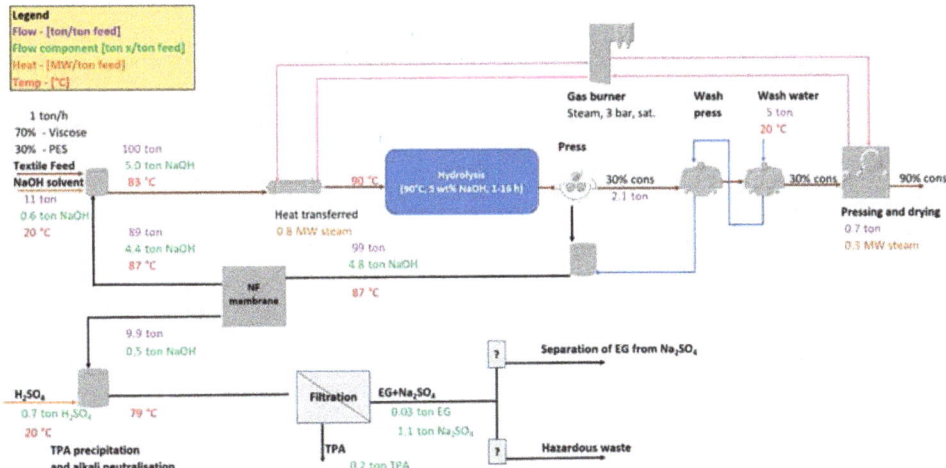

Figure 8. The proposed process layout for the viscose/PET separation process with the improvement of separating and recycling residual alkali back to the textile feed. Purple values indicate total flow, green indicates component flow in the stream, brown indicates heat transfer, and red indicates the temperature. The values are to be seen as indicative and, in this case, refer to a hydrolysis residence time of 16 h.

It is of great importance to allow the PET hydrolysis to be completed to avoid contamination of the viscose fraction by the remaining PET. Experimental results indicate that up to 16 h are required for complete hydrolysis of neat PET and the Filament70/30 sample. However, for the blended fabric, Fabric70/30, complete removal of PET was not achieved, even after 24h of hydrolysis (Figure 2a). Phase transfer catalysts have previously been used in alkaline hydrolysis of both PET and cotton/PET blends, and their use might be advisable to speed up reaction, debottleneck the process, and minimize the volume of the reaction vessel. One potential alternative would be to conduct the hydrolysis in multiple step reactors, with a countercurrent flow of the alkaline solution.

The proposed process in this work depicts a standalone plant. It is well known that integration between chemical industrial processes might have various economic and environmental benefits, as compared to a standalone process. Integration of the viscose/PET separation process with a Kraft pulp mill would yield several potential benefits:

1. Instead of having a gas burner, low pressure steam could be bought from the pulp mill. This removes the capital expenditure (gas burner) of the viscose/PET process, and, further, the pulp mill steam should be less expensive, as well as renewable, compared to natural gas.
2. If it is not possible to separate the EG/Na_2SO_4 fraction, the existing infrastructure at the pulp mill may offer some alternatives for its disposal. If the EG/Na_2SO_4 stream can be treated as a normal effluent, it could simply be sent to the wastewater treatment plant. Otherwise, it could perhaps be evaporated, and subsequently incinerated in the recovery boiler; however, this would be at the expense of increased energy consumption in the mill's evaporation plant. Furthermore, it must be investigated whether the amount of Na_2SO_4 in the stream will significantly alter the pulp mill's Na/S balance, possibly increasing the NaOH makeup demand, as well as the associated costs.

3. The chemicals used in the viscose/PET process, NaOH and H_2SO_4, are already used in the pulp mill. The viscose/PET process can potentially directly use the prepared solutions of these chemicals from the pulp mill.
4. Other economic benefits related to infrastructure, such as a reduced need for land and ground preparation, shared buildings, and utilities infrastructure (e.g., power substation, fresh- and cooling water), as well as shipping and/or transportation.

4. Conclusions

Alkaline hydrolysis of PET from viscose/PET fabrics was shown to render pure terephthalic acid and a cellulose fraction with a decreased molecular weight and increased crystallinity. A complete depolymerization of PET, as well as the subsequent separation of the blended fabric, requires long reaction times, >24 h under the present, mild reaction conditions. The intrinsic viscosity of cellulose after hydrolysis is 130–150 mL g^{-1}, too low for use in commercially regenerated fibers or traditional cellulose derivatives, such as cellulose ethers or nitrates. In fact, all samples show an initial IV, prior to hydrolysis, below the desired IV range for dissolving pulps aimed for use in various commercial uses, probably due to a reduction in molecular weight as a function of viscose processing. However, low molecular weight polysaccharides can find other uses, e.g., through hydrolysis into nanocrystalline cellulose, or in sugar platforms for further valorization into specialty chemicals and biofuels. An upscaled process layout was constructed, based on the proposed process, with the objective of optimizing chemical and energy use. From this model, a number of process developments were suggested. We believe that an iterative process between small-scale experimental work and techno–economic and life cycle analysis is highly beneficial in developing recycling processes which are economically and ecologically viable.

Future Suggestions

The inefficient use of NaOH was recognized as the weakest point in the proposed process, whose cost, at present, would very likely exceed any reasonable revenue from the sales of TPA, EG, and the cellulose fraction. Efforts must be dedicated to reducing NaOH consumption, e.g., by the use of an NF membrane or a lower textile-to-solvent ratio. These results should be used as a guide to further experimental work, to favor an economically and ecologically viable process.

Further, experimental work should focus on increasing the hydrolysis rate, as the slow kinetics of hydrolysis of the blended textile require large volume reactors. This may be accomplished using a phase transfer catalyst, or an increase in either temperature or NaOH concentration.

In future studies, proper characterization of the EG/Na_2SO_4 stream and identification of separation technologies and waste treatment alternatives should be included. Given the low content of EG in the stream, it is possibly more relevant to identify technologies which could be used to separate Na_2SO_4, which dominates the composition of the stream. Furthermore, the process would probably largely benefit from an integration with a pulp mill.

It is emphasized that the values, presented as output from the process layouts, should be seen as indicative for the overall process. Even though some aspects of the proposed process do not seem encouraging at this current stage, it is far too early to reject it. Instead, the suggested actions to take in future work should be seen as important steppingstones to close the loop for a fraction of the multi-material textiles.

Author Contributions: Conceptualization: J.B., M.B. and K.J.; Data curation, A.P., J.W. and A.I.; Funding acquisition, H.d.l.M.; Investigation, A.P., J.W. and A.I.; Writing—original draft, A.P., J.W. and A.I.; Writing—review and editing, J.B., M.B., K.J. and H.d.l.M. All authors have read and agreed to the published version of the manuscript.

Funding: This research was funded by Södra Skogsägarnas stiftelse för Forskning, Utveckling och Utbildning, grant number 2019-106.

Acknowledgments: Södras forskningsstiftelse is gratefully acknowledged for financial support, and Södra Innovation för fruitful discussions. Jehona Sjöberg, Angéle Cruz, Carina Berglund, and Elham Franzén are acknowledged for their invaluable experimental help. We would also like to acknowledge the use of the Chalmers Material Characterization Lab (CMAL), and the help from Michal Strach.

Conflicts of Interest: The authors declare no conflict of interest. The industrial reference group contributed to the formulation of the research question, but had no role in the design, execution, interpretation, or writing of the study.

References

1. The Fiber Year GmbH. *The Fiber Year 2019*; The Fiber Year GmbH: Speicher, Switzerland, 2019.
2. Piribauer, B.; Bartl, A. Textile recycling processes, state of the art and current developments: A mini review. *Waste Manag. Res.* **2019**, *37*, 112–119. [CrossRef] [PubMed]
3. Haslinger, S.; Hummel, M.; Anghelescu-Hakala, A.; Määttänen, M.; Sixta, H. Upcycling of cotton polyester blended textile waste to new man-made cellulose fibers. *Waste Manag.* **2019**, *97*, 88–96. [CrossRef] [PubMed]
4. De Silva, R.; Wang, X.; Byrne, N. Recycling textiles: The use of ionic liquids in the separation of cotton polyester blends. *RSC Adv.* **2014**, *4*, 29094–29098. [CrossRef]
5. Baghaei, B.; Compiet, S.; Skrifvars, M. Mechanical properties of all-cellulose composites from end-of-life textiles. *J. Polym. Res.* **2020**, *27*, 260. [CrossRef]
6. Ouchi, A.; Toida, T.; Kumaresan, S.; Ando, W.; Kato, J. A new methodology to recycle polyester from fabric blends with cellulose. *Cellulose* **2010**, *17*, 215–222. [CrossRef]
7. Piribauer, B.; Bartl, A.; Ipsmiller, W. Enzymatic textile recycling—Best practices and outlook. *Waste Manag. Res.* **2021**, *39*, 1277–1290. [CrossRef]
8. Sinha, V.; Patel, M.R.; Patel, J.V. PET waste management by chemical recycling: A review. *J. Polym. Environ.* **2010**, *18*, 8–25. [CrossRef]
9. Palme, A.; Peterson, A.; de la Motte, H.; Theliander, H.; Brelid, H. Development of an efficient route for combined recycling of PET and cotton from mixed fabrics. *Text. Cloth. Sustain.* **2017**, *3*, 4. [CrossRef]
10. Brelid, H.; Bogren, J. A Process for Separation of the Cellulosic Part from a Polyester and Cellulose Composition. U.S. Patent WO2020013755A1, 2 September 2021.
11. Glaus, M.A.; Van Loon, L.R. Degradation of cellulose under alkaline conditions: New insights from a 12 years degradation study. *Environ. Sci. Technol.* **2008**, *42*, 2906–2911. [CrossRef]
12. Rollins, M.L.; Tripp, V.W. Optical and electron microscopic studies of cotton fiber structure. *Text. Res. J.* **1954**, *24*, 345–357. [CrossRef]
13. Broadbent, A.D. Artificially made fibres based on cellulose. In *Basic Principles of Textile Coloration*; Sociaty of Dyers and Colourists: Bradford, UK, 2001; pp. 92–106.
14. Luo, M.; Roscelli, V.A.; Neogi, A.N.; Sealey, J.E., II; Jewell, R.A. Lyocell Fibers, and Compositions for Making the Same. U.S. Patent CA2323437C, 9 July 2001.
15. Sixta, H.; Potthast, A.; Krotschek, A.W. Chemical pulping processes: Sections 4.1–4.2.5. In *Handbook of Pulp*; WILEY-VCH Verlag GmbH & Co. KGaA: Weinheim, Germany, 2006; pp. 109–229.
16. Sandin, G.; Peters, G.M. Environmental impact of textile reuse and recycling—A review. *J. Clean. Prod.* **2018**, *184*, 353–365. [CrossRef]
17. Östlund, Å.; Wedin, H.; Bolin, L.; Berlin, J. *Textilåtervinning—Tekniska Möjligheter och Utmaningar*; Naturvårdsverket: Stockholm, Swedish, 2015.
18. Vanzetto, A.B.; Beltrami, L.V.R.; Zattera, A.J. Textile waste as precursors in nanocrystalline cellulose synthesis. *Cellulose* **2021**, *28*, 6967–6981. [CrossRef]
19. Ruiz-Caldas, M.X.; Carlsson, J.; Sadiktsis, I.; Jaworski, A.; Nilsson, U.; Mathew, A. Cellulose Nanocrystals from Postconsumer Cotton and Blended Fabrics: A Study on Their Properties, Chemical Composition, and Process Efficiency. *ACS Sustain. Chem. Eng.* **2022**, *10*, 3787–3798. [CrossRef]
20. Sanchis-Sebastiá, M.; Ruuth, E.; Stigsson, L.; Galbe, M.; Wallberg, O. Novel sustainable alternatives for the fashion industry: A method of chemically recycling waste textiles via acid hydrolysis. *Waste Manag.* **2021**, *121*, 248–254. [CrossRef] [PubMed]
21. Sofokleous, M.; Christofi, A.; Malamis, D.; Mai, S.; Barampouti, E.M. Bioethanol and biogas production: An alternative valorisation pathway for green waste. *Chemosphere* **2022**, *296*, 133970. [CrossRef]
22. Sanchis-Sebastiá, M.; Novy, V.; Stigsson, L.; Galbe, M.; Wallberg, O. Towards circular fashion—Transforming pulp mills into hubs for textile recycling. *RSC Adv.* **2021**, *11*, 12321–12329. [CrossRef]
23. Immergut, E.H.; Schurz, J.; Mark, H. Viskositätszahl-Molekulargewichts-Beziehung für Cellulose und Untersuchungen von Nitrocellulose in verschiedenen Lösungsmitteln. *Mon. Chem. Verwandte Teile And. Wiss.* **1953**, *84*, 219–249. [CrossRef]

24. Borysiak, S.; Garbarczyk, J. Applying the WAXS method to estimate the supermolecular structure of cellulose fibres after mercerisation. *Fibres Text. East. Eur.* **2003**, *11*, 104–106.
25. Knill, C.J.; Kennedy, J.F. Degradation of cellulose under alkaline conditions. *Carbohydr. Polym.* **2002**, *51*, 281–300. [CrossRef]
26. Kumar, S.; Guria, C. Alkaline hydrolysis of waste poly(ethylene terephthalate): A modified shrinking core model. *J. Macromol. Sci. Part A* **2005**, *42*, 237–251. [CrossRef]
27. Wennerström, M.; Bylund, S. Method for Controlling Viscosity in Dissolving Pulps. AI Patent WO 2017/105322, 22 June 2017.
28. Prado, K.S.; Gonzales, D.; Spinacé, M.A.S. Recycling of viscose yarn waste through one-step extraction of nanocellulose. *Int. J. Biol. Macromol.* **2019**, *136*, 729–737. [CrossRef] [PubMed]
29. European Commision. From the Sugar Platform to Biofuels and Biochemicals. 2015. Available online: https://ec.europa.eu/energy/sites/ener/files/documents/EC%20Sugar%20Platform%20final%20report.pdf (accessed on 23 January 2022).
30. BISFA. Terminology of Man-Made Fibres. 2017. Available online: http://www.bisfa.org/wp-content/uploads/2018/06/2017-BISFA-Terminology-final.pdf (accessed on 19 April 2022).
31. Wagman, D.D.; Evans, W.H.; Parker, V.B.; Schumm, R.H.; Halow, I.; Bailey, S.M.; Churney, K.L. The NBS tables of chemical thermodynamic properties. Selected values for inorganic and C_1 and C_2 organic substances in SI units. *J. Phys. Chem. Ref. Data* **1982**, *11*, 38, 58, 232–299.
32. Ruiz, J.; Rincón, C.; Contreras, L.; Sidney, R.R.; Almarza, C. Sustainable and negative carbon footprint solid-based NaOH technology for CO_2 capture. *ACS Sustain. Chem. Eng.* **2020**, *8*, 19003–19012. [CrossRef]
33. Koch. Available online: Https://www.kochseparation.com/wp-content/uploads/2020/10/Selro-NF-MPS-34-2-5-and-4-inch-elements.pdf (accessed on 15 February 2022).
34. Recyctec. Available online: https://www.recyctec.se/en/ (accessed on 15 February 2022).
35. Stena Recycling. Available online: https://www.stenarecycling.se/en/ (accessed on 15 February 2022).

Article

Investigation of the Physical Properties of Yarn Produced from Textile Waste by Optimizing Their Proportions

Hafeezullah Memon [1,*], Henock Solomon Ayele [2], Hanur Meku Yesuf [2,3] and Li Sun [4]

[1] College of Textile Science and Engineering, International Institute of Silk, Zhejiang Sci-Tech University, Hangzhou 310018, China
[2] Ethiopian Institute of Textile and Fashion Technology, Bahir-Dar University, Bahir-Dar 1037, Ethiopia; henockso11@gmail.com (H.S.A.); 419005@mail.dhu.edu.cn (H.M.Y.)
[3] College of Textiles, Donghua University, Shanghai 201620, China
[4] Hangzhou Pulay Information Technology Co., Ltd., Hangzhou 310016, China; sunli@mypulay.com
* Correspondence: hm@zstu.edu.cn

Abstract: Since textile waste recycling is a global challenge, there is an emerging need to explore this research direction due to the little knowledge about textile recycling. This study aimed to study the property of yarns produced from recycled textile/cotton fiber blends for proportion optimization and to check whether they can be used for denim fabric production. The properties of recycled fiber and virgin cotton spun on open-end having 4.5 Ne were investigated with fiber proportions of 20/80, 25/75, 30/70, 35/65, 40/60, 45/55, and 50/50. The results were analyzed with Design-Expert software, using central composite design to optimize the proportion. The 40/60 proportion had the optimum result, and by using this optimized proportion, 10 Ne yarn was produced and used for denim fabric production. The sample denim fabric produced used recycled yarn as a weft, showing that the recycled fiber turned yarn can be used in manufacturing products such as denim. The physical properties of the denim fabric confirmed that the recycled goods have wearable quality. Since this research can be applied on an industrial scale, it would benefit textile academia, industry, the environment, and society.

Keywords: textile waste; virgin cotton; denim fabric; Design-Expert software

1. Introduction

Textile waste is categorized into pre-consumer waste and post-consumer waste [1]; the former is generated in the manufacturing process, and the latter contains worn-out or trashed textile products that are no longer serviceable [2]. Among the most severe environmental problems, solid textile waste management is the most significant one society faces. It has been well addressed that waste incineration and improper waste management in landfills have harmful environmental consequences. Some sustainable clothing consumption behaviors [3] and textile upgrading [4] have been proposed to solve this problem. Even though some textile wastes can be used as a fuel in waste to energy processes, this has a higher tendency to increase CO_2 emissions; in that regard, incineration of waste is a much better option comparably given that there are no options [5,6]. Additionally, chemical recycling has been considered even worse than producing new virgin materials [7,8]. Since synthetic textile fabrics do not decompose, thus creating substantial environmental consequences, they produce methane gas, contributing to global warming after decomposition. Furthermore, textiles in landfills also release hazardous substances into soil and groundwater, contributing to environmental pollution. In the denim industry, mostly cotton is used, and several works have been conducted in the past to process denim [9–11]. Thus, a small percentage of pre- and post-consumer waste is reprocessed into fibers that can be used as yarn for woven or knitted fabrics [12]. This process of recycling textile fabrics comes with both environmental and economic benefits. Additionally, it immensely reduces the space required for landfills; has the benefit of not requiring energy- and time-intensive pre-processes for virgin materials; and reduces the costs of dyeing, scoring, fixing agents, and water consumption for their manufacturing [13,14].

Pre-consumer wastes are the industrial wastes generated in fibrous products' manufacturing processes. Since wastes are generated at different manufacturing stages, the type of waste could be a single polymer or a very complex multi-material compound. Single polymers are usually useful, more manageable, and suitable for recycling as a corporate recycling process can separate them to their corresponding component [15]. Such approaches help identify the source of waste generation and design measures to reduce waste generation [16,17].

Post-consumer waste, also known as home waste, refers to fibrous materials and by-products scrapped after their service life is over [18,19]. Since most fibers are transformed into different products, the volume of post-consumer waste is higher when compared with the fiber consumption rate [20]. Some collection and sorting network chains are required to support commercially available activities to recycle this home waste significantly. In addition to directly reusing clothing, the collected textile fiber waste could be converted into usable nonwoven applications such as wipes or shredded into being used as fillers [21]. In addition, recently, the microstructure and performance characteristics of acoustic insulation materials made from post-consumer recycled denim fabrics have been assessed [11,22–24].

Sharma and Goel [25] developed nonwoven fabrics using polyester/cotton recycled fibers with different proportions (70:30, 50:50, and 70:30) with the needle punching method, and the various properties of the developed nonwoven fabrics were analyzed. It was indicated that the most proper ratio of nonwoven blended fabric based on physical properties was the 30/70 cotton/polyester fiber blend. Ichim and Sava [26] also studied the spinnability of recycled cotton fibers blended with virgin cotton at 20/80, 40/60, and 60/40 proportions. The result revealed that as the percentage of waste increases, the yarn has decreased breaking elongation but increased yarn irregularity and mass irregularity. Additionally, they indicated that these types of yarns with the percentage of reclaimed cotton could be used as a gauze bandage fabric due to its short lifespan and because their application focuses on absorbency rather than uniformity and strength. Thilak and Dhandapani [27] studied the influence of blend ratio on the quality aspects of recycled cotton and polyester yarn and indicated that the ratio of recycled polyester significantly affects the quality of recycled cotton and polyester blended yarn. As a result of increasing the tenacity of the blended yarn, the elongation at break and hairiness increases, and correspondingly, it decreases unevenness, thick and thin places, and neps.

In contrast, decreasing the linear density increases the tenacity, thick and thin places, elongation at break, neps, unevenness, and hairiness [27,28]. The intrinsic nature of the recycling process produces fibers with a short length of unopened or partially opened fibers, non-uniform, and a higher number of imperfections [29]. These restrictions enable only the production of coarser count yarns. However, the production of medium and fine count yarns is yet to be explored. Wei and Liang found that properly using waste textiles and clothing recycling has gained much attention in the post-COVID era [30]. New spinning systems, including friction, rotor, and ring spinning, have successfully produced recycled polyester and cotton yarns. Rotor spinning is one of the most widely used spinning techniques for producing reclaimed yarns [31].

Recently, Wang et al. [32] studied virgin polyester and cotton blending using principal component analysis and grey relational analysis. This work investigated the properties of recycled yarn produced from recycled fiber and virgin cotton with different proportions. The recycled fiber and virgin cotton were mixed in different proportions to coarser count yarns. Producing finer yarn is a future task that needs further research based on the results we acquire from this research. The number of runs and proportions were identified using Design-Expert software's central composite design method [33].

2. Experimental

2.1. Materials

Pre-consumer (industrial) knitted waste fabric was collected from a cutting and stitching department of the garment industry. When cutting knitted fabric for garment production, much of the knitted fabric becomes waste material due to improper pattern laying and cutting problems. In addition, pre-consumer knitted waste fabric occurs in the knitting process due to defects. All defective knitted fabric was removed during fabric inspection. The type of waste and other waste fiber and fabric parameters were measured and are presented in Tables 1 and 2. The waste comprised different colors and parameters; therefore, it is taken in a range.

Table 1. Pre-consumer waste fabric parameters.

Fabric Type	Knitting
Structure	Plain
Weight (gm/m^2)	140–180
Thickness (mm)	0.4–0.6
Course density (cpi)	40–50
Wale density (wpi)	30–40
Yarn count (tex)	18–30
Yarn twist (tpm)	150–230

Table 2. Parameters of waste acrylic weft yarn.

Fiber Type	Regenerated Acrylic
Yarn count (tex)	738
Yarn twist (tpm)	185
Breaking strength (N)	24.6
Elongation (%)	11.97

Waste of weft yarn (recycled acrylic) was used as an additional material in the recycled fiber. This was mixed with waste fabric, and both were converted to fiber as shown in Figure 1; the parameters of waste yarn are shown in Table 2.

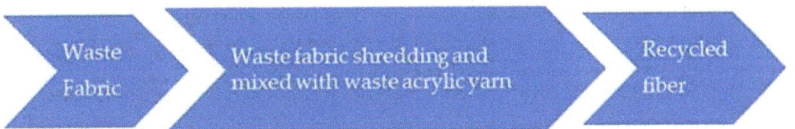

Figure 1. Steps to recycle waste fabrics.

The collected fabric and yarn wastes were mixed and converted into recycled fiber using a Garnet shredding machine, and the recycled fiber was then tested for the different properties as shown below in Table 3. The fiber length, fineness, and strength were measured. The trash content was analyzed by the Shirley trash analyzer. Chemical (solvent) identification showed that cotton, acrylic, and polyester fibers were found in the recycled fiber at percentages of 32.5%, 49.8%, and 15.9%, respectively. The remaining percentage could be protein fibers such as wool and silk. The chemical analysis was carried out as per the ASTM-D5103-07 standard by using sulfuric acid (99.8%), sodium hydroxide (pellets), and zinc chloride.

Table 3. Recycled fiber properties.

Parameters	Value
Fiber length (mm)	30
Fiber fineness (dtex)	1.75
Fiber strength (cN/tex)	27.6
Elongation (%)	13.5
Trash content and noil (%)	12.42

Virgin cotton fiber was collected from the factory, and the properties were tested using a High Volume Instrument; the results are presented in Table 4.

Table 4. Virgin cotton fiber properties.

Parameters	Unit	Average Value	Standard	Performance Level
Moisture	%	6.3	4.5–6.5	Low moisture
Micronaire value	-	3.58	3.0–3.6	Fine
Maturity	-	0.82	0.75–0.85	Immature
UHML	mm	26.4	26.2–27.8	Medium-long
Uniformity index	%	80.8	81–84	Medium
Short fiber content	%	9.4	10–13	Medium
Strength	g/tex	25.1	25–27	Medium
Elongation	%	8.4	≥ 7	Very high
Trash content	%	6.26	4–7	Dirty
Sticky point	No.	0.6	0–2	Very small

2.2. Yarn Production

The recycled fiber and virgin cotton were mixed in different proportions. The number of runs and proportions were identified using the central composite design method from Design-Expert software [34,35], as shown in Table 5.

Table 5. Optimized central composite design for recycled fiber/cotton fiber proportion.

Std	Run	Factor 1 A: Recycled Fiber%	Factor 1 B: Recycled Fiber%
2	1	50	50
6	2	35	65
5	3	25	75
4	4	40	60
3	5	20	80
1	6	30	70

Yarn of 131 tex (4.5 Ne) was manufactured on a rotor spinning machine with seven different combinations of recycled fiber and cotton fiber. The yarn was investigated for its physical properties. The result was inserted into the software and analyzed using different tools, as presented in Table 6.

Table 6. Two-level factorial design for recycled fiber/cotton fiber proportion.

Std	Run	Factor 1 A: Recycled Fiber%	Factor 1 B: Recycled Fiber%
5	1	35	65
4	2	40	60
6	3	45	55
7	4	30	70
2	5	20	80
3	6	25	75
1	7	50	50

This software performed another hundred runs between those seven proportions depending on the given results. From the 100 runs, the proportions which had the optimum result were selected. The selected optimum proportion was used to produce a 10 Ne count of yarn, and sample denim fabric was produced using this yarn. Thus, the fabric used recycled fiber/cotton blended yarn as a weft yarn because it was challenging to produce in a warp.

2.3. Denim Fabric Production

During sample denim fabric production, the recycled yarn is used as weft yarn because denim fabric is made of dyed warp and undyed weft yarn, so the recycled yarn cannot be used as warp; it cannot be dyed. The second reason is that there is the insufficient yarn for the warping process. In addition, it is better to use recycled yarn as a weft thread as it is hiding in the back side of the fabric. However, there is high weft breakage because the twist was not in the required amount. It has a twist multiplier of less than 3, but the required twist multiplier is 5.2–5.6. If it produces with the required twist level, it can be used for denim fabric production rather than a blanket. The fabric was produced using a TOYOTA air jet weaving machine with a loom speed of 848 rpm.

3. Results and Discussion

In order to study the effect of recycled fiber and cotton fiber content on the yarn property, seven different proportions were produced and are presented in Figure 2 from the software.

Figure 2. Yarns produced with different proportions of recycled fiber and cotton fiber.

The analyzed results by Design-Expert software with two factors (Factor 1: Recycled fiber; Factor 2: Cotton fiber) are presented in Table 7. The six responses (Response 1: Yarn strength; Response 2: Yarn elongation; Response 3: Yarn unevenness; Response 4: Thin place; Response 5: Thick place; Response 6: Neps) are presented in Tables 8 and 9.

Table 7. Denim fabric construction and weaving process parameters.

Parameters	Results
Warp yarn	Cotton
Weft yarn	Recycled/cotton blend
Warp and weft yarn count	10 s open-end yarn
Warp density	60 end per inch
Weft density	42 pick per inch
Weave type	3/1 right-hand twill

Table 8. Description of factors (recycled and cotton fiber).

Factors	Fiber	Unit	Minimum	Maximum	Mean	Std dev
A	Recycled	%	20.00	50.00	35.00	10.80
B	Cotton	%	50.00	80.00	65.00	10.80

Table 9. Description of responses (yarn properties).

Response	Property	Unit	Obs	Max	Min	Mean	Std dev
R1	Yarn strength	cN/tex	7	7.13	4.38	5.95	0.8597
R2	Yarn elongation	%	7	10.85	6.35	9.29	1.67
R3	Yarn unevenness	U%	7	15.64	12.11	13.57	1.16
R4	Thin place (−50%)	-	7	68	12	33.14	17.81
R5	Thick place (+50%)	-	7	812	104	345.5	251.58
R6	Neps (+200%)	-	7	852	198	390.4	226.88

The six basic properties (Responses) of yarn were studied, and the result is presented in Table 9.

Predicted value vs. actual value: The predicted value was compared with the actual value for each response. In Figure 3, the solid line indicates the predicted value, and the dots indicate the actual value. The actual values of yarn strength, elongation, and unevenness coincided with the predicted values, as shown in Figure 3.

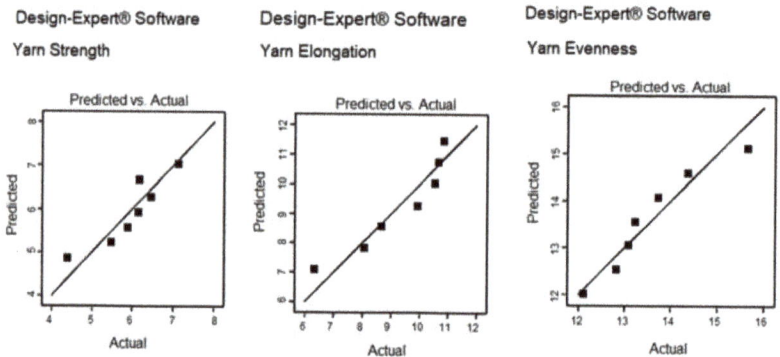

Figure 3. Predicted vs. actual values of yarn strength, elongation, and unevenness.

However, the actual values of yarn imperfection (neps as well as a thin and thick places) did not coincide with the predicted values as shown in Figure 4, which means that for this model, the yarn imperfections were less affected by the recycled fiber or cotton fiber.

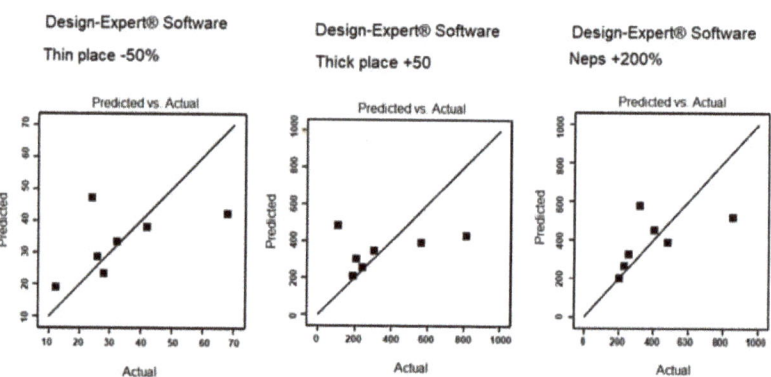

Figure 4. Predicted vs. actual values of neps and thin and thick places.

3.1. Correlation and Regression

The relation between each factor and response is shown below in Figure 5: as the color goes to blue, the relation is a strong negative correlation (inverse relation), and as the color goes to red, the relation is a strong positive correlation (direct relation).

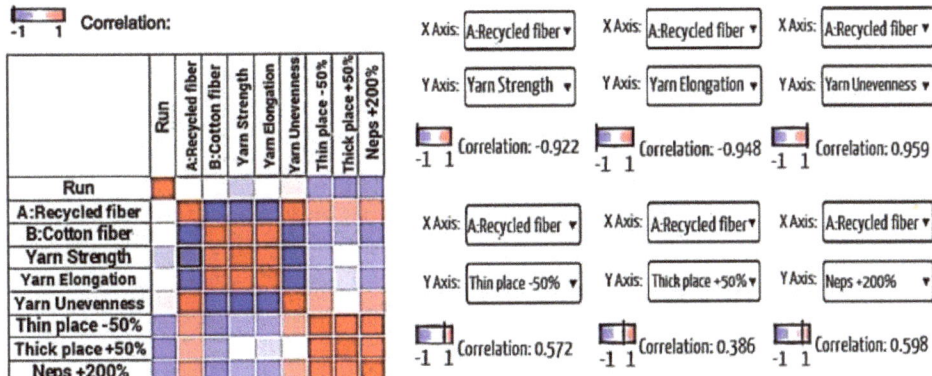

Figure 5. Correlation between recycled fiber and yarn properties.

Recycled fiber has a strong negative correlation with yarn strength and yarn elongation (−0.922 and −0.948, respectively), which means that as recycled fiber increases, the yarn strength and elongation decrease. But it has a strong positive correlation with yarn unevenness, 0.959. It correlates 0.572, 0.386, and 0.598 with the thin place, thick place, and neps, respectively. The correlation between cotton fiber and each response is inverse of the correlation between recycled fiber and responses. Cotton fiber has a strong positive correlation with yarn strength and yarn elongation (0.922 and 0.948, respectively), which means that as cotton fiber increases, the yarn strength and elongation also increase. But it has a strong negative correlation with yarn unevenness, −0.959; as the cotton fiber increase, the yarn unevenness will decrease. The cotton fiber has a correlation of −0.572, −0.386, and −0.598 with thin and thick places and neps, respectively, shown in Figure 6. Therefore, yarn strength, elongation, and unevenness are highly affected by the proportions of cotton and recycled fiber, and the yarn-produced imperfections (thin and thick places and neps) are affected to some extent.

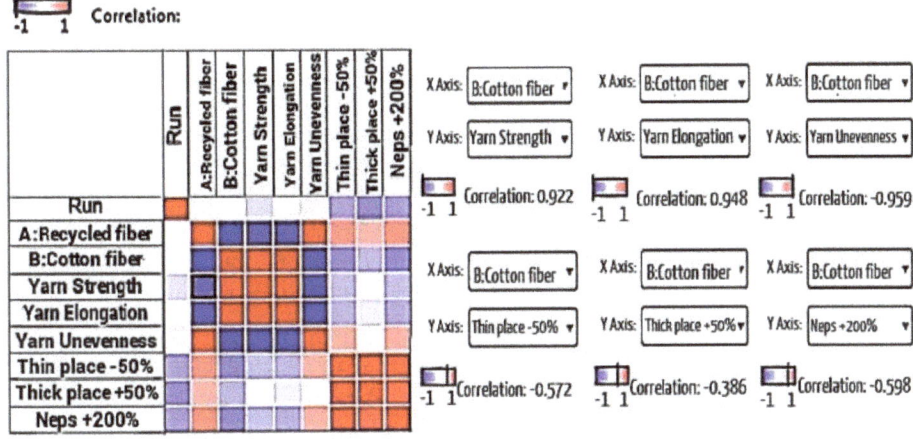

Figure 6. Correlation between cotton fiber and yarn properties.

3.2. Interaction between Factor and Response

3.2.1. Interaction between Cotton Fiber/Recycled Fiber Proportion and Yarn Strength

The interaction between the factors (cotton fiber/recycled fiber proportion) and Response 1 (yarn strength) indicates that as the cotton fiber percentage increases, the yarn strength increases, but as the recycled fiber amount increases, the yarn strength decreases as shown in Figure 7. The optimum result was observed with the proportion of 72% cotton fiber and 28% recycled fiber.

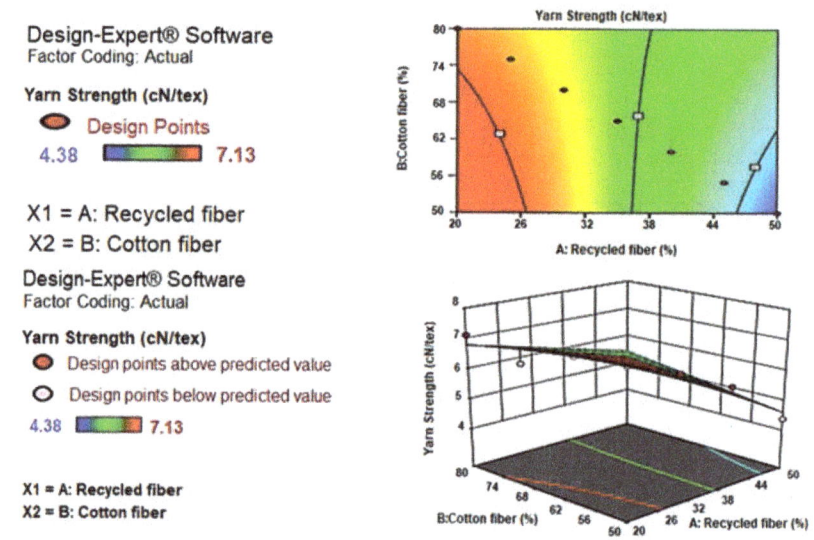

Figure 7. Two- and three-dimensional plots of the interaction of cotton fiber/recycled fiber proportion and yarn strength.

3.2.2. Interaction between Cotton Fiber/Recycled Fiber Proportion and Yarn Elongation

The interaction between the factors (cotton fiber/recycled fiber proportion) and Response 2 (yarn elongation) indicated that as the cotton fiber percentage increases, the yarn elongation also increases, but as the recycled fiber amount increases, the yarn elongation decreases, as shown in Figure 8. The optimum result was observed with 66% cotton fiber and 24% recycled fiber, indicated by the red color.

3.2.3. Interaction between Cotton Fiber/Recycled Fiber Proportion and Yarn Unevenness

The interaction between the factors (cotton fiber/recycled fiber proportion) and Response 3 (yarn unevenness) indicated that as the cotton fiber percentage increases, the yarn unevenness decreases. However, as the recycled fiber amount increases, the yarn unevenness increases, as shown in Figure 9. To obtain the minimum result, cotton fiber should be above 70% and recycled fiber should be below 26%, which is indicated by the blue color.

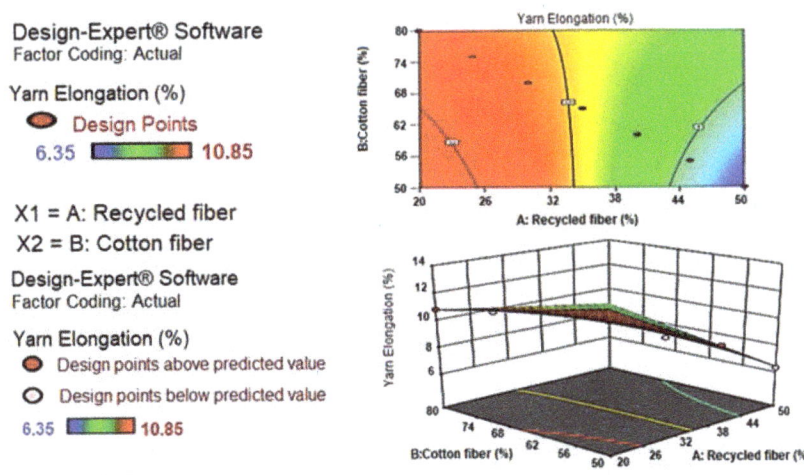

Figure 8. Two- and three-dimensional plots of the interaction of cotton fiber/recycled fiber proportion and yarn elongation.

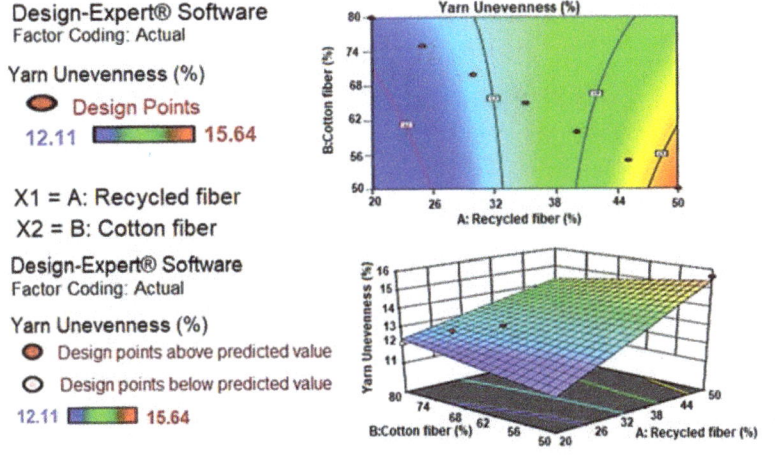

Figure 9. Two- and three-dimensional plots of the interaction of cotton fiber/recycled fiber proportion and yarn unevenness.

3.2.4. Interaction between Cotton Fiber/Recycled Fiber Proportion and Thin Place (−50%)

The interaction between the factors (cotton fiber or recycled fiber proportion) and Response 4 (thin place) indicated that as the cotton fiber percentage increases, thin places on the yarn decrease, but as the recycled fiber amount increases, thin places increase, as shown in Figure 10. Thus, to obtain a favorable result, cotton fiber should be above 70% and recycled fiber should be below 26%, which is indicated by the blue color.

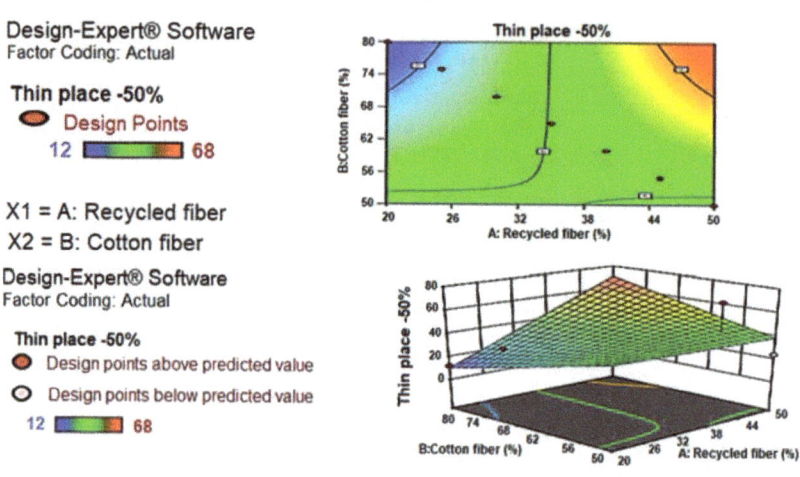

Figure 10. Two- and three-dimensional plots of the interaction of cotton fiber/recycled fiber proportion and yarn thin place (−50%).

3.2.5. Interaction between Cotton Fiber/Recycled Fiber Proportion and Thick Place (+50%)

The interaction between the factors (cotton fiber/recycled fiber proportion) and Response 5 (thick place) indicated that as the cotton fiber percentage increases, thick places on the yarn decrease, but as the recycled fiber amount increases, thick places increase, as shown in Figure 11. However, sometimes there is also a sudden change, possibly due to another factor. To obtain a favorable result, cotton fiber should be above 71% and recycled fiber should be below 25%, which is indicated by the blue color.

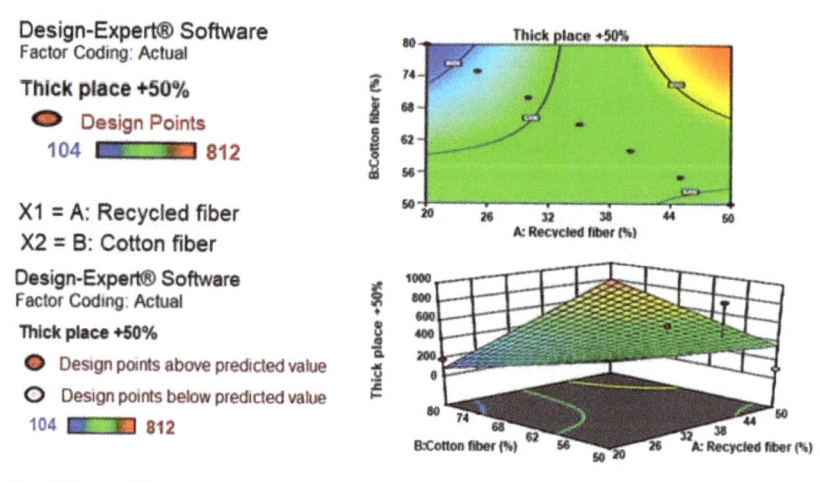

Figure 11. Two- and three-dimensional plots of the interaction of cotton fiber/recycled fiber proportion and yarn thick place (+50%).

3.2.6. Interaction between Cotton Fiber/Recycled Fiber Proportion and Neps (+200%)

The interaction between the factors (cotton fiber/recycled fiber proportion) and Response 6 (Neps) indicated that as the cotton fiber percentage increases, neps on the yarn decrease, but as the recycled fiber amount increases, neps also increase, as shown in Figure 12. To obtain a favorable result, cotton fiber should be above 60% and recycled fiber should be below 28%, which is indicated by the blue color.

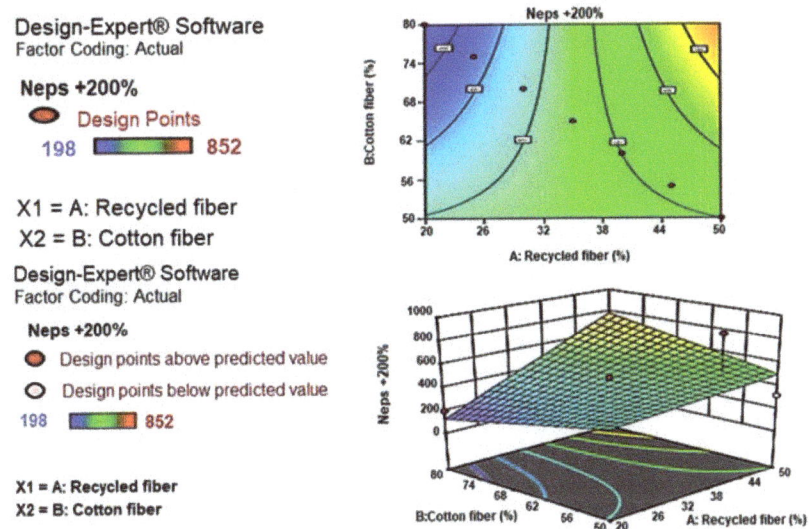

Figure 12. Two- and three-dimensional plots of the interaction of cotton fiber/recycled fiber proportion and Neps (+200).

After the responses were analyzed, one hundred runs were made by the software to find out the optimum proportion, as shown in Figure 13. The yarn properties were predicted for each of the hundred runs, and then, the optimum result was selected (the first run in Figure 13).

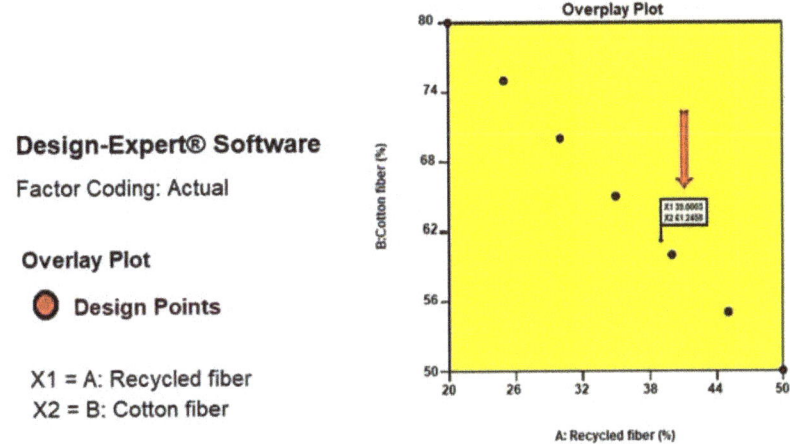

Figure 13. Graphical selection of optimum proportion of recycled and cotton fiber.

Finally, the software confirmed the optimum proportion, 39/61, with 95% confidence, as shown in Table 10. Thus, this proportion was used to produce the yarn of 10 Ne count for sample denim fabric production (as shown in Figure 14).

Table 10. Test results of each response for seven runs.

Trial	F1 Recycled fiber (%)	F2 Cotton fiber (%)	R1 Yarn strength (cN/tex)	R2 Yarn elongation (%)	R3 U%	R4 Thin place (−50%)	R5 Thick place (+50%)	R6 Nep (+200%)
1	20	80	6.13	9.89	13.23	32	307	480
2	25	75	5.9	8.67	13.74	42	562	404
3	30	70	5.46	8.07	14.37	68	812	852
4	35	65	6.44	10.54	13.07	26	212	254
5	40	60	7.13	10.85	12.11	12	184	198
6	45	55	6.2	10.63	12.81	28	238	227
7	50	50	4.38	6.35	15.64	24	104	318

Figure 14. Denim fabric produced from recycled yarn and cotton yarn.

Some physical properties of the produced sample fabric, such as tensile strength, elongation, tear strength, pilling, and abrasion, were tested, and the result is presented in Table 11.

Table 11. Confirmation of the selected optimum proportion.

	Confirmation Two-Sided, Confidence = 95% Recycled Fiber = 39.0003, Cotton Fiber = 61.2458	
Response	Prediction Mean	Prediction Median
Yarn strength	5.65512	5.65512
Yarn elongation	8.70025	8.70025
Yarn unevenness	13.978	13.978
Thin place (−50)	36.9145	36.9145
Thick place (+50)	381.517	381.517
Neps (+200)	440.718	440.718

The result shows that the fabric has favorable tensile strength and elongation, even if it was produced without spandex (Table 12). The abrasion and pilling test indicated a moderate change in color with visual fraying and slight to moderate pilling. However, further study is needed to analyze denim fabric produced from recycled and pure cotton yarn.

Table 12. Physical properties of sample denim fabric.

Parameters	Results
Thickness (mm)	1.07
Tensile strength (N)	429, in weft direction
Elongation (%)	11.25, in weft direction
Tear strength (N)	48.62, in the weft direction
Pilling	3/4
Abrasion	3
Warp and weft yarn count	10 s OE

4. Conclusions

As per the designed method, the results show that as the composition of recycled fiber in the blend increases, the yarn properties such as strength, elongation, and evenness are reduced, and vice versa. The yarn imperfections increased to some extent as the recycled fiber amount increased. Thus, yarn strength, yarn elongation, and yarn evenness are significantly affected by the recycled fiber/cotton fiber proportion. Therefore, from the seven proportions (20/80, 25/75, 30/70, 35/65, 40/60, 45/55, and 50/50 recycled fiber/cotton fiber) and from the other 100 proportions between those seven proportions, which were made by Design-Expert software, the optimum result was found with the 39/61~40/60 proportion. Additionally, 10 Ne yarn was produced using this optimized proportion and used for denim fabric production, even if it was not on the required specification. The sample fabric had favorable physical properties. Therefore, the recycled fiber can be used to produce denim fabric by producing yarn with the required specification.

Author Contributions: Conceptualization, H.M. and H.S.A.; methodology, H.M. and H.S.A.; software, H.M. and H.S.A.; validation, H.M. and H.M.Y.; formal analysis, H.M. and H.S.A.; investigation, H.S.A. and H.M.Y; resources, H.M. and H.M.Y.; data curation, H.M. and H.M.Y.; writing—original draft preparation, H.M. and H.S.A.; writing—review and editing, H.M. and L.S.; visualization, H.M., H.M.Y. and L.S.; supervision, H.M.; project administration, H.M.Y.; funding acquisition, H.M. All authors have read and agreed to the published version of the manuscript.

Funding: This work was supported by the Research Fund for International Scientists (RFIS-52150410416), the National Natural Science Foundation of China, and the Research Startup grant of ZSTU (20202294-Y).

Institutional Review Board Statement: Not applicable.

Informed Consent Statement: Not applicable.

Data Availability Statement: Not applicable.

Acknowledgments: We acknowledge the technical staff of Eitex, Bahir-Dar University, for assisting us in obtaining trails and experiments.

Conflicts of Interest: The authors declare no conflict of interest. In addition, the funders had no role in the design of the study; in the collection, analysis, or interpretation of data; in the writing of the manuscript, or in the decision to publish the results.

References

1. Haslinger, S.; Hietala, S.; Hummel, M.; Maunu, S.L.; Sixta, H. Solid-state NMR method for the quantification of cellulose and polyester in textile blends. *Carbohydr. Polym.* **2019**, *207*, 11–16. [CrossRef]
2. Li, X.; Wang, L.; Ding, X. Textile supply chain waste management in China. *J. Clean. Prod.* **2021**, *289*, 125147. [CrossRef]
3. Jianfang, L.; Wanying, C. Research on status and dilemma of sustainable clothing consumption behavior. *J. Silk* **2020**, *57*, 18–25.

4. Liu, A.; Guo, J. Cycle fashion: Research on the development and design method of textile and garment upgrading. *J. Silk* **2020**, *57*, 132–139.
5. Beshah, D.A.; Tiruye, G.A.; Mekonnen, Y.S. Characterization and recycling of textile sludge for energy-efficient brick production in Ethiopia. *Environ. Sci. Pollut. Res.* **2021**, *28*, 16272–16281. [CrossRef]
6. Sandin, G.; Peters, G.M. Environmental impact of textile reuse and recycling—A review. *J. Clean. Prod.* **2018**, *184*, 353–365. [CrossRef]
7. Weiran, Q.; Pinghua, X.; Laili, W. Review on Polyester Fiber Recycling and Progress of Its Environmental Impact Assessment. *Adv. Textile Technol.* **2021**, *29*, 22–26. [CrossRef]
8. Qian, W.; Ji, X.; Xu, P.; Wang, L. Carbon footprint and water footprint assessment of virgin and recycled polyester textiles. *Text. Res. J.* **2021**, *91*, 2468–2475. [CrossRef]
9. Abid, S.; Hussain, T.; Nazir, A.; Raza, Z.A.; Siddique, A.; Azeem, A.; Riaz, S. Simultaneous Fixation of Wrinkle-Free Finish and Reactive Dye on Cotton Using Response Surface Methodology. *Cloth. Text. Res. J.* **2017**, *36*, 119–132. [CrossRef]
10. Ahmad, S.; Ashraf, M.; Abid, S.; Jabbar, M.; Shafiq, F.; Siddique, A. Recent Developments in Laser Fading of Denim: A Critical Review. *J. Nat. Fibers* **2022**, 1–11. [CrossRef]
11. Siddique, A.; Hassan, T.; Abid, S.; Ashraf, M.; Hussain, A.; Shafiq, F.; Khan, M.Q.; Kim, I.S. The Effect of Softeners Applications on Moisture Management Properties of Polyester/Cotton Blended Sandwich Weft-Knitted Fabric Structure. *Coatings* **2021**, *11*, 575. [CrossRef]
12. Aronsson, J.; Persson, A. Tearing of post-consumer cotton T-shirts and jeans of varying degree of wear. *J. Eng. Fiber Fabr.* **2020**, *15*, 1558925020901322. [CrossRef]
13. Grebosz-Krawczyk, M.; Siuda, D. Attitudes of young European consumers toward recycling campaigns of textile companies. *Autex Res. J.* **2019**, *19*, 394–399. [CrossRef]
14. Tshifularo, C.A.; Patnaik, A. Recycling of plastics into textile raw materials and products. In *Sustainable Technologies for Fashion and Textiles*; Elsevier: Amsterdam, The Netherlands, 2020; pp. 311–326.
15. Wang, H.; Memon, H.; Abro, R.; Shah, A. Sustainable Approach for Mélange Yarn Manufacturers by Recycling Dyed Fibre Waste. *Fibres Text. East. Eur.* **2020**, *3*, 18–22. [CrossRef]
16. Chaka, K.T. Beneficiation of Textile Spinning Waste: Production of Nonwoven Materials. *J. Nat. Fibers* **2021**, 1–10. [CrossRef]
17. Bizuneh, B.; Tadesse, R. Investigation of Ethiopian apparel industry's fabric waste and its management strategies. *J. Text. Inst.* **2022**, *113*, 141–150. [CrossRef]
18. Dash, A.K.; Nayak, R. Management of protective clothing waste. In *Waste Management in the Fashion and Textile Industries*; Elsevier: Amsterdam, The Netherlands, 2021; pp. 233–251.
19. Jiaqi, L. Research on the Design of Used Clothes Recycling Space Oriented by Experiential Business. *Furnit. Inter. Des.* **2020**, *9*, 116–119. [CrossRef]
20. Stanescu, M.D. State of the art of post-consumer textile waste upcycling to reach the zero waste milestone. *Environ. Sci. Pollut. Res.* **2021**, *28*, 14253–14270. [CrossRef]
21. Leal Filho, W.; Ellams, D.; Han, S.; Tyler, D.; Boiten, V.J.; Paço, A.; Moora, H.; Balogun, A.-L. A review of the socio-economic advantages of textile recycling. *J. Clean. Prod.* **2019**, *218*, 10–20. [CrossRef]
22. Islam, S.; El Messiry, M.; Sikdar, P.P.; Seylar, J.; Bhat, G. Microstructure and performance characteristics of acoustic insulation materials from post-consumer recycled denim fabrics. *J. Ind. Text.* **2020**, 1528083720940746. [CrossRef]
23. Siddique, A.; Hussain, T.; Ibrahim, W.; Raza, Z.A.; Abid, S. Optimization of discharge printing of indigo denim using potassium permanganate via response surface regression. *Pigm. Resin Technol.* **2018**, *47*, 228–235. [CrossRef]
24. Siddique, A.; Hussain, T.; Ibrahim, W.; Raza, Z.A.; Abid, S.; Nazir, A. Response surface optimization in discharge printing of denim using potassium permanganate as oxidative agent. *Cloth. Text. Res. J.* **2017**, *35*, 204–214. [CrossRef]
25. Sharma, R.; Goel, A. Development of nonwoven fabric from recycled fibers. *J. Text. Sci. Eng.* **2017**, *7*, 289–292.
26. Ichim, M.; Sava, C. Study on recycling cotton fabric scraps into yarns. *Bul. Agir.* **2016**, *3*, 65–68.
27. Vadicherla, T.; Saravanan, D. Effect of blend ratio on the quality characteristics of recycled polyester/cotton blended ring spun yarn. *Fibres Text. East. Eur.* **2017**, *25*, 48–52. [CrossRef]
28. Memon, H.; Khoso, A.N.; Memon, S. Effect of dyeing parameters on physical properties of fibers and yarns. *Int. J. Appl. Sci. Eng. Res.* **2015**, *4*, 401–407.
29. Wagaye, B.T.; Adamu, B.F.; Jhatial, A.K. Recycled Cotton Fibers for Melange Yarn Manufacturing. In *Cotton Science and Processing Technology: Gene, Ginning, Garment and Green Recycling*; Wang, H., Memon, H., Eds.; Springer: Singapore, 2020; pp. 529–546.
30. Shansen, W.; Jianfang, L. The impact of COVID-19 on the attention to sustainable clothing consumption:Analysis of Baidu indexes on old clothes recycling, old clothes renovation and old clothes donation. *J. Silk* **2021**, *58*, 40–46. [CrossRef]
31. Shi, J.; Liang, W.; Wang, H.; Memon, H. Recent Advancements in Cotton Spinning Machineries. In *Cotton Science and Processing Technology: Gene, Ginning, Garment and Green Recycling*; Wang, H., Memon, H., Eds.; Springer: Singapore, 2020; pp. 165–190.
32. Hanqi, W.; Weili, W.; Xuefeng, G.; Weilai, C. Grey Relational Analysis of Principal Components of Sustainable Cotton Blended Yarn. *Adv. Text. Technol/* **2021**, *29*, 55–61. [CrossRef]
33. Memon, H.; Chaklie, E.B.; Yesuf, H.M.; Zhu, C. Study on Effect of Leather Rigidity and Thickness on Drapability of Sheep Garment Leather. *Materials* **2021**, *14*, 4553. [CrossRef]

34. Xingying, Z.; Xiaojun, Y.; Dongpo, G. Influence of Vibration Characteristic Parameters on Vibration Response of Single Degree of Freedom System. *Pack. Eng.* **2020**, *41*, 75–81. [CrossRef]
35. Pengcheng, Z.; Junyuan, W.; Linyu, M.; Yiming, C. Optimization of process parameters for preparation of CS/PEO by electrospinning based on response surface methodology. *J. Silk* **2020**, *57*, 31–34. [CrossRef]

MDPI
St. Alban-Anlage 66
4052 Basel
Switzerland
Tel. +41 61 683 77 34
Fax +41 61 302 89 18
www.mdpi.com

Sustainability Editorial Office
E-mail: sustainability@mdpi.com
www.mdpi.com/journal/sustainability

www.ingramcontent.com/pod-product-compliance
Lightning Source LLC
LaVergne TN
LVHW070728100526
838202LV00013B/1190